FAST ETHERNET IMPLEMENTATION AND MIGRATION SOLUTIONS

Fast Ethernet Implementation and Migration Solutions

Martin Nemzow

McGraw-Hill, Inc.
New York • San Francisco • Washington, D.C. •
Auckland • Bogotá • Caracas • Lisbon • London •
Madrid • Mexico City • Milan • Montreal • New Delhi •
San Juan • Singapore • Sydney • Tokyo • Toronto

Library of Congress Cataloging-in-Publication Data

Nemzow, Martin A. W.
Fast Ethernet implementation and migration : solutions / Martin Nemzow.
p. cm.—(McGraw-Hill series on computer communications)
Includes index.
ISBN 0-07-046385-9 (pb)
1. Ethernet (Local area network system) I. Title. II. Series.
TK5105.8.E83N463 1997 97-6919
004.6'8—dc21 CIP

McGraw-Hill

A Division of The **McGraw-Hill** *Companies*

1 2 3 4 5 6 7 8 9 0 FGR/FGR 9 0 2 1 0 9 8 7

ISBN 0-07-046385-9

The sponsoring editor for this book was Steve Elliot, the editing supervisor was Scott Amerman, and the production supervisor was Suzanne Rapcavage. It was set in Vendome ICG by Joanne Morbit of McGraw-Hill's Professional Book Group composition unit, Hightstown, N.J.

Printed and bound by Quebecor/Fairfield.

CONTENTS

Contents

PREFACE

If your network is strapped for bandwidth, failing under its own success, and in desperate need of a makeover to address loading issues, performance bottlenecks, and new multimedia and Intranet applications, you need this book to provide the business strategies and implementational technologies and help you make the leap to Fast Ethernet.

Fast Ethernet represents an incremental migration from the original Ethernet network. For the purpose of this book, Fast Ethernet represents all switched and special-purpose Ethernet technology. This includes switched Ethernet, duplexed Ethernet, as well as switched and duplexed Ethernet running at the standard 10-Mbps transmission rate. In addition, a variant of Ethernet called IEEE 802.9 now includes ISDN bundled into the standard Ethernet delivery signal, a useful technology for standard desktop network with Internet integration and either analog or digital desktop videoconferencing.

Fast Ethernet is more readily recognized as Ethernet packet delivery at 100 Mbps. However, this faster delivery method has been broken into two camps with competing standards. The primary standard is called 100Base-T and is represented by a new and enhanced cabling method represented by Category 5 UTP wire. The second standard is called 100BaseVG-AnyLAN that works on standard 10Base-T which is made up of four pairs wired to every socket and patch cord. While 100BaseVG-AnyLAN is interoperable with standard Ethernet through bridges, routers, or special hubs, this technology does not use the standard collision-based overload protection method of Ethernet.

Instead, VG-AnyLAN inserts a hub-based demand priority (token-passing) scheme for increased reliability for mission-critical networking. In addition, the AnyLAN protocol provides a migration path both for standard Ethernet at 10 Mbps as well as for 4 Mbps and 16 Mbps Token-Ring networks. A booming market of Fast Ethernet hubs, cards, and duplexed and switched solutions enhances your options and the likelihood of success when you are designing a strategy to migrate to Fast Ethernet. In addition, new fast Level 2 switching and faster Level 3 routing in enterprise equipment provide more bandwidth and fast delivery times for both workgroup, enterprise, Intranet, and Internet Service Provider performance bottlenecks. Gigabit Ethernet is slowly entering the market, providing even faster Ethernet migration options for server, workstation, and backbone requirements; in fact, they may eventually enable display of desktop and workstation implementations of Fibre Channel and ATM. While these "fast Ethernet" solutions are viable and cost-effective, details make or break the implementation, the reason for *Fast Ethernet Implementation and Migration Solutions.*

Before you commit to a faster Ethernet strategy, you must understand the ramifications of hardware, software, design, and performance choices, technologies, and implementation issues. This book addresses physical wiring issues, the technical information needed to choose between two-pair and four-pair Fast Ethernet standards, between the competing Fast Ethernet implementations, and all the hidden stumbling blocks. As solutions go, Fast Ethernet is straightforward, obvious, and desirable. As demand for network hardware increases, the implementational costs will orphan the standard 10Base-2, 10Base5, and 10Base-T solutions. While Ethernet 10/100 Mbps may provide a migration option for some time to come, Fast Ethernet will become the backbone and desktop connectivity solution of choice and Gigabit Ethernet will support server-to-server and high-bandwidth backbone applications.

Acknowledgements

A hearty a thank you to the vendors who provided software, hardware, and know-it-all to build faster networks and troubleshoot them. Special thanks to Steven Elliot, acquisition editor at McGraw-Hill. Here are some fast packets to Carol Weingrod, who mediated in big ways the progress of this project and ministered in smaller ways to the scheduling activities of this book. I want to thank Sophie Esther and Gabriel Charles for stringing cable, drilling holes through beams and CBC walls, terminating many RJ-45 connections, and otherwise overseeing the physical side of network construction and maintenance.

FAST ETHERNET IMPLEMENTATION AND MIGRATION SOLUTIONS

CHAPTER 1

An Overview

"Fast Ethernet" is, strictly speaking, a direct enhancement of 802.3 Ethernet. The primary references are IEEE 802.3u or 100Base-T. However, Fast Ethernet also represents any addition to an Ethernet network to improve its basic performance. Faster Ethernets include switched 10-Mbps and switched 100-Mbps Ethernet, duplexed Ethernet, switched and duplexed Ethernet, 100Base-T, 100Base-TX, 100Base-T4, 100Base-X, 100Base-F, and 100BaseVG-AnyLAN (IEEE 802.12). One vendor is even selling daisy-chained 155-Mbps LAN equipment. The formal committee for Ethernet is already beginning talks for Gigabit Ethernet for multicast video and desktop videoconferencing, an extension of Ethernet already dubbed "Ultra Fast Ethernet."

It is clear why this explosion for faster protocols is occurring. First, Ethernet is the basis for more than 75% of all installed LANs and enterprise networks. Second, most strategic applications (that is, integrated workflow programming solutions) require greater bandwidth and reduced transmission delays. Third, the technology is available for Fast Ethernet. The reality is that Fast Ethernet is cheap and it is fast. When the network load burden warrants the costs for fast half-duplex, full-duplexed, and switched migration, it is worthwhile.

Unfortunately, however, "Fast Ethernet" is not one thing, one definition, nor a uniform implementation. The confusion is not because Ethernet is per se a confusing standard, but rather Fast Ethernet has developed ahead of any formal standards. In fact, early adopters of the original Ethernet discovered similar disparities with Ethernet Type I, Type II, Type III, 802.3, Synoptics Trellis, and differences in heartbeat and packet type formats. While those birth pains are all but gone, Fast Ethernet represents a new maturity with some frankly adolescent rebellion. Two formal Fast Ethernet standard families are represented by 100Base-T and 100BaseVG. Interoperability is not assured among different vendors, while product differences create many connectivity and functionality concerns. Multiple possibilities for cabling adds to the general confusion.

Fast Ethernet is presented as the likely successor to LANs and enterprise networks overloaded by bandwidth. Such migration approaches often misfire, unless the network is carefully designed and intelligently administered. There are subtle limitations and caveats with any migration strategy, and Fast Ethernet is no exception. Networks are complex in all phases, from design to organization, execution, daily operation, and maintenance. A well-run network can match the promise of Fast Ethernet and give an organization a powerful strategy that outstrips the competition.

This book provides the practical knowledge needed to design, build, grow, and maintain a Fast Ethernet communication network for most vendor equipment, including 3Com, Bay Networks, Cabletron, Cisco, DEC, Hewlett-Packard, Kalpana, and many others. Issues such as interconnecting networks to form enterprise networks and successfully achieving interoperability are addressed, as are the cost and benefits achieved through host mainframe downsizing and installation of client/server networks.

The Purpose of This Book

Fast Ethernet Implementation and Migration Solutions is designed primarily to answer questions facing an Ethernet LAN or enterprise network team plotting network redesign and upgrades to the wiring infrastructure. The specifications of the IEEE 802 committee and the ISO network standard are used as the basis for definition of key terms, because most vendors try to adhere to these standards. Much of the material in this book relates to physical wiring, cabling, and connectivity issues, if only because the success of 100-Mbps transmission hinges on the quality of the physical cabling plant. Although physical issues seem to be "beneath" the threshold of most network managers, initial lack of attention to cabling and plant issues with Fast Ethernet will just raise the issues later. This book discusses EIA/TIA wiring quality standards and suggestions, and the ramifications of Category 5 certification and technical service bulletin (TSB) 67.

Fast Ethernet Implementation and Migration Solutions includes many floor plans, schematic drawings, comparisons of physical versus logical network designs, sample project management planning sessions, discussions of strategic planning, and performance planning. Fast Ethernet is not the only fast network infrastructure, and in fact, is often not the best migration strategy. Asynchronous Transmission Mode (ATM), Frame Relay, Fiber Distributed Data Interface, Fibre Channel, and Enterprise backplanes provide alternatives worth reviewing as substitutes for (or partial solutions in conjunction with) Fast Ethernet.

Vendor-supplied Ethernet documentation is supplemented and explained in this book, and its practical ramifications are discussed. In addition, the nuts and bolts of planning for installation and cabling, capacity planning, and physical maintenance are presented for the busy network administrator. Concise and specific illustrations and descriptions prepare even the novice for all stages of network administration, to clarify the ambiguous and sparse vendor documentation for Fast Ethernet and Category 3, Category 5, and fiber cabling.

Although magazines have published many articles about networking with Fast Ethernet, the material is usually focused to benchmark performance of ultimate throughputs, and is thus very limited in depth. Few books are available on this topic; most exist as specialized technical reference manuals from the vendor or third-party publishers. Furthermore, periodical literature is geared to what is new and what is better, and thus does not address the migration requirements of 10-year-old network infrastructures. Consider also the prosaic daily grind of maintaining a fully operational information highway while trying to replace it with one at least 10 times faster. This book targets these gaps to provide information on design, planning, installation, enhancement and upgrades, management, performance, and troubleshooting. In effect, *Fast Ethernet Implementation and Migration Solutions* delivers insight into other ways—some obvious, most practical—to guide you through a massive infrastructure transition.

Intended Audience

The intended audience includes MIS directors, industry consultants, network managers, workgroup managers, software developers building client/server and distributed networking applications, support technicians, computer professionals, and anyone who uses a network or is involved in network management, operations, design, planning, maintenance, and implementation. Anyone who is intending to migrate an existing Ethernet network will find this material valuable for its insight and performance caveats. Anyone redesigning the work environment, specifically with regard to networking work flow, can benefit from the strategic information and practical know-how. This book is also a useful teaching tool; it covers many network environments from the vantage point of specific technical details, but also with the perspective of general problem-solving methods.

Book Content

The content of *Fast Ethernet Implementation and Migration Solutions* is straightforward. This book provides some case studies for faster networks' requirements and transitions, gives a brief history on Ethernet, defines networking protocols, and shows you how to design, install, test, manage, and debug Ethernet networks.

Specifically, haphazard migration or implementation of Fast Ethernet is not likely to provide benefits. Instead, you could create reliability and new performance problems. The Category 3 and Category 5 cable definitions in the TIA/EIA/BISCI, ANSI/IEEE/ECMA standards are good in concept but often are not completely implemented through ignorance, lack of attention to detail, poor installation, and lack of tools to actually certify them. Furthermore, poor network design is likely to increase bandwidth (one of the ideas behind Fast Ethernet) but at the expense of latency, a killer of client/server, PowerBuilder or Visual Basic networked applications, video-on-demand, teleconferencing, and other time-sensitive applications. While this is an inherent limitation of switching and translational hubs, it can also be avoided with careful design.

This book is designed to provide abundant information for all aspects of Fast Ethernet operations, for both small office networks and enormous enterprise networks. I have tried to include practical experience from building large internetworks, debugging clients' problems, tuning overloaded networks, and dealing with the day-to-day grind on sparsely used networks. The book contains references to legacy networks because 60 percent of all Ethernet networks contain fewer than 50 nodes and have been installed for more than 5 years, and most are "sparse"; and it is these networks that are the grist for integration and migration projects. The other 40 percent of networks are stressed by traffic saturation, routing problems, growth pains, and interconnectivity and interoperability demands; this represents a morass of bandwidth and latency trade-offs not addressed so directly by migrations to Fast Ethernet.

The Structure of This Book

This section presents the organization of *Fast Ethernet Implementation and Migration Solutions.* As an acknowledgment of the reader's limited time, Figure 1.1 illustrates the design and flow of the knowledge contained within this book.

Chapter 1 is this overview of the book explaining the purpose, audience, book content, and structure of the book.

Chapter 2 is a concise explanation of standard Ethernet and the various Fast Ethernet technologies. A lot of material may be also be presented

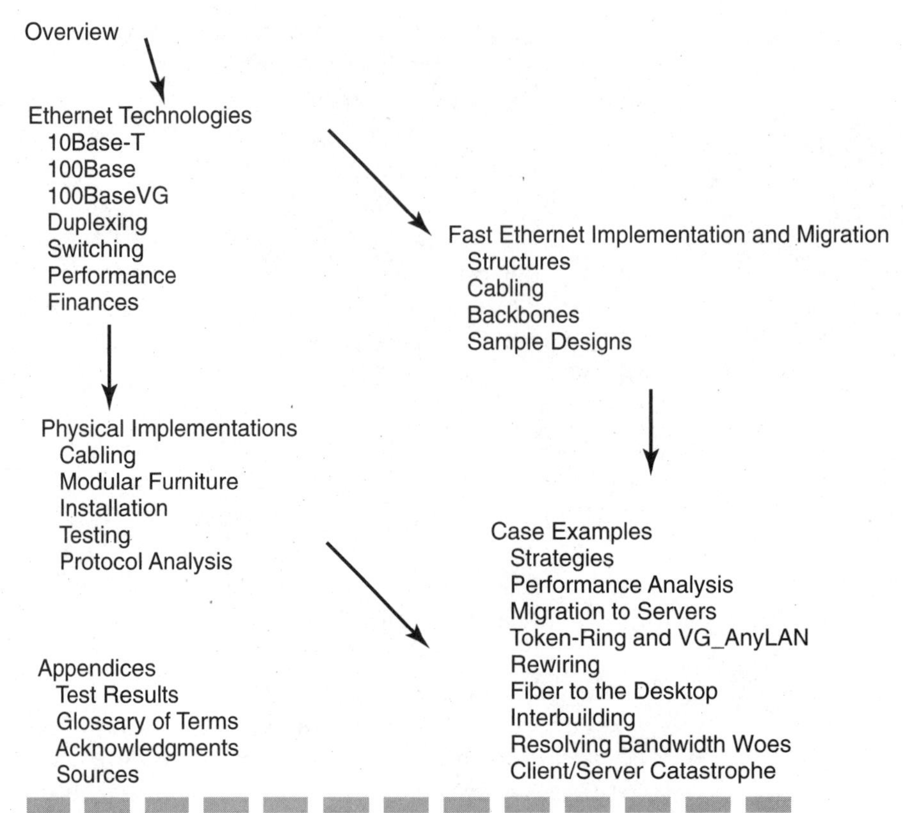

Figure 1.1.
The organization of *Fast Ethernet Implementation and Migration Solutions.*

in tables. This includes speeds, materials and infrastructure, limitations, and the parts needed to put it all together. This chapter will also address switching and duplexed connections.

Chapter 3 presents a verbal and very graphical view of network designs. This will show floor plans, schematic wiring plans, and renditions of copper and fiber wiring plants. Because Ethernet (particularly Fast Ethernet) is unlikely to exist in a vacuum or become the material to totally replace older Token-Ring, ARCNET, or Ethernet, there will be liberal references to older-generation networks, wireless, Frame Relay, ISDN, ATM, and FDDI backbones.

Chapter 4 expands on the 2-pair, 4-pair, coax and fiber wiring and encoding models, and explains variations on pair usage and encoding schemes. This chapter discusses the effects of these on installation, management, and costs, and will go into the nitty-gritty of patching wires, making connectors, and dealing with architecture issues related to installation. This chapter also will show how to use the new tools for testing, and how to test vLAN and switched setups (a real problem) when the hubs do not support SNMP, SNMP-II, or RMON, or provide a generalized network-management platform.

Chapter 5 shows real-world implementations with vendor names, user sites with names, diagrams, litanies of problems, design issues, insights into the decision-making process, and what went right or wrong. The hardest part of this chapter, and the material everyone really wants, is *what went wrong* and *for whom* and *why,* so that they can avoid these same problems.

The appendix includes reference to cited materials and sources for statistics, models, software, notes, standards, and photographs. There is also a glossary of terms specific to Fast Ethernet, defined and cross-referenced by the all-too-common acronyms.

CHAPTER 2

Ethernet Technologies

Ethernet is strictly defined by a single packet technology and transmission methodology at the media level. It is described by a frame format and transmission-signaling protocol at all logical network levels above the hardware. This chapter defines what Ethernet is and what all the Fast Ethernet variants are relative to the standard Ethernet and to each other. The variants have different limitations that you need to know about if you are going to create a successful migration from a slow network to a fast one. You do not want to get stuck with a collection of many fast components that, in aggregate as a network, perform only as well as (or even worse than) the original network.

Success with Fast Ethernet is about design, interoperability, and recognition of the goals for the network upgrade and migration. Consider also the meaning of the word "migration" as a long, slow journey. While it is possible to convert a slow network to Fast Ethernet in massive network redeployment, most changes will represent analyzed and incremental improvments to an existing network infrastructure. Although this book does not always explore the technical details of Ethernet or Fast Ethernet, it does show the overall picture of what you need to know to implement a successful Fast Ethernet migration strategy. My other books present the details, trivia, and managerial and debugging techniques for Ethernet, Token-Ring, and FDDI network management. That information is a supplement if you need or want it. This book is focused on the transition from one architecture to a faster one. *Fast Ethernet Implementation and Migration Solutions* gives you the details and opinions so that you can make effective decisions.

For example, the detail of the Ethernet frame format is summarized by four components: address, type, data, and error-detection fields, as shown in Figure 2.1.

The transmission methodology—called the *protocol*—is best described as a crowd moderated by the concept that only a single (one-way) conversation can occur at one time. Initial Ethernet conversation speed was set at 10,000,000 bits per second (10 Mbps) when it was first introduced in 1977. Interestingly, AT&T sold the first commercial "Ethernet" products using standard telephone wire and a transmission bandwidth of only 1 Mbps. The product was called *StarLAN.* This book is about everything above the 10-Mbps speed, and methods for migrating to faster speeds while maintaining compatibility with the initial and subsequent protocol variants. Although IsoEthernet (IEEE 802.9a) is the bundling of standard 10Base-T 10-Mbps Ethernet with four ISDN BRI lines for operation on standard Category 3 UTP, it is an *enhancement* of Ethernet for support of data communications and videoconferencing or integration with

Frame element	Byte location
Address	1 to 12
Destination	1 to 6
Source	7 to 12
Type or length field	13 to 14
Data field	15 to 1515 (variable)
FCS (CRC)	Last 4 bytes

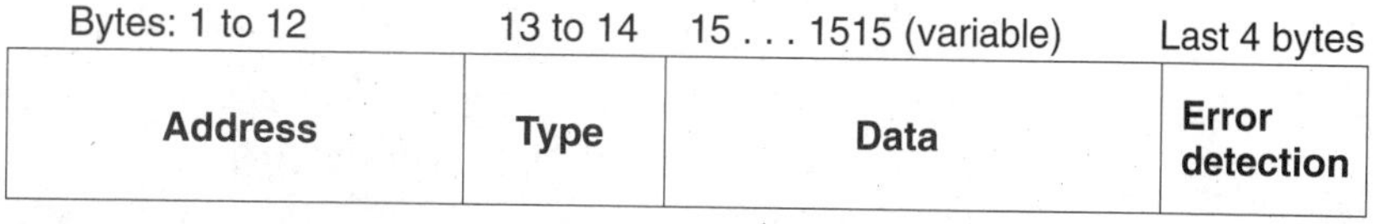

Bytes: 1 to 12	13 to 14	15 . . . 1515 (variable)	Last 4 bytes
Address	Type	Data	Error detection

Figure 2.1.
The Ethernet frame with address, type, data, and error-detection components. This frame is the same for all Fast Ethernet variants.

telecommunication. Although IsoEthernet provides 10 plus 6.144 Mbps of bandwidth, it is really not a faster Ethernet or a *Fast Ethernet,* just standard Ethernet with some extra channels. Hence, for the purposes of this book, Fast Ethernet includes duplexed Ethernet, switched-10 Ethernet, 100-Mbps Ethernet, and switched-100 Ethernet.

In Ethernet, two or more overlapping conversations is a shouting match that is solved by a "time-out" and a randomized waiting period. The time-out is called an *Ethernet collision,* and the waiting period is called the transmission *backoff.* The backoff waiting period increases with subsequent collision counts. Note that collisions are not errors per se, but rather are a normal part of the protocol. Collisions reduce available bandwidth and decrease data transmission capacity. Duplexed and switched Ethernet disable the collision mechanism, while 100BaseVG bypasses it entirely with a round-robin methodology.

Although taking turns (as with Token-Ring, FDDI, *Priority Access Control Enabled* [PACE] in 100BaseVG, or demand priority) seems more orderly and efficient, the statistical reality is that Ethernet is more *efficient* and *faster* than prioritized or tokenized protocol schemes. Delivery time for Ethernet is faster than Token-Ring, and delivery time for Fast Ethernet is faster than FDDI. This does not mean that Ethernet is better than token-based protocols, or that it is a direct replacement for Token-Ring and FDDI. Networks with high traffic levels or sustained transaction processing loads often require deterministic token-based delivery. This is particularly true with mission-critical client/server-based software applications. Performance is a function both of available bandwidth and the transmission and protocol methods. Analysis of performance requirements requires a complex statistical review of application loads, network area coverage, server response times, and user expectations. Fast Ethernet represents some diverse technologies which in general might not be completely appropriate to your migration requirements.

Standard Ethernet or Fast Ethernet alone does not provide the robustness of a Token-Ring or FDDI, unless combined with full duplex, switching, or flattened enterprise hub architectures. Although the non-deterministic Ethernet protocol is more efficient than token priority management, there are two flaws to this seemingly chaotic and messy protocol. First, efficiency is lost when the Ethernet network traffic load is pushed above a critical saturation level. At saturation, more traffic pushed onto the network just means that less actual work is accomplished. Saturation can result in total traffic collapse. The critical saturation level is a nonlinear function of the traffic load based on packet sizes and transmission request frequencies. On most Ethernet LANs, this saturation occurs when utilization reaches about 30 percent, as shown in Figure 2.2.

It doesn't take very much to create a 10-Mbps Ethernet with saturation problems. A network doesn't need to have a sustained 30-percent

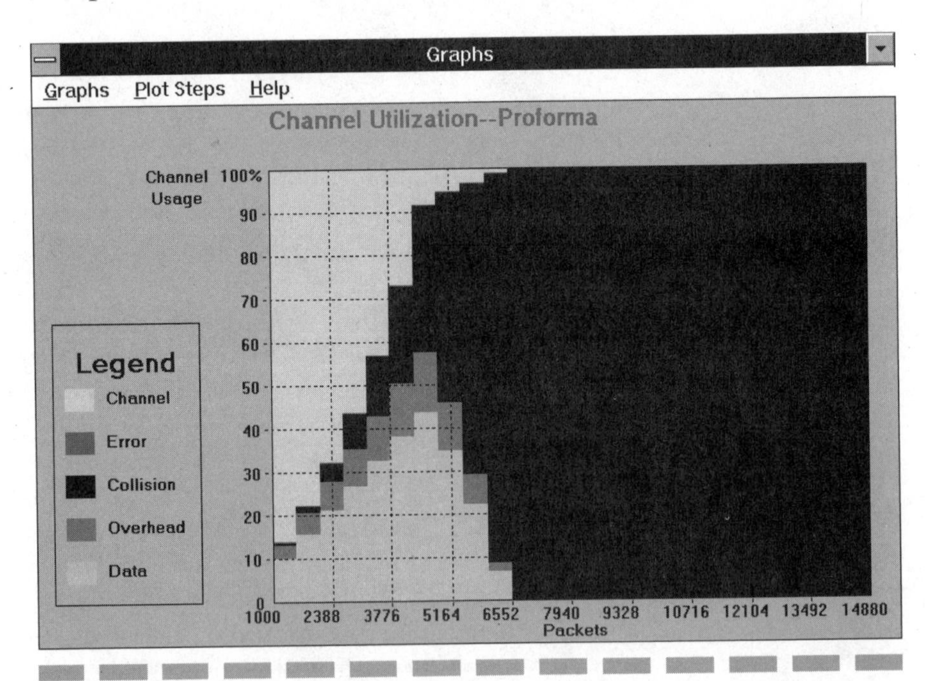

Figure 2.2.
Network saturation occurs when collisions crowd out data traffic. (Courtesy of Network Performance Institute.)

load to create a performance bottleneck, just a few peaks of a few seconds' duration every once in a while to cause a sustained traffic jam. Client/server applications do it easily because of the traffic loads, the many bidirectional messages between client and server, and the congestion on the network channel.

Second, the backoff algorithm as originally implemented is flawed. When saturation occurs, priority *defaults* to those network devices least needing that priority. The network devices most needing to transmit, which are primarily servers and hosts responding to client requests, are squeezed out. Although there is a formal proposal to fix the flawed binomial backoff algorithm with a logarithmic fix, the entrenched reality of 80 million preexisting network interface cards with the flawed algorithm in hardware or firmware enforces the status quo. Just as the Dvorak keyboard is unlikely to displace QWERTY, the Ethernet standard (with the flawed collision-handling algorithm) will indubitably endure as the entrenched networking standard. Despite these two defects, some of which can be easily bypassed, Fast Ethernet is a strategic enhancement to the Ethernet standard. The future migration path includes full-duplexing and switching, which alone prevent collisions.

Any conformity to networking standards represents important progress towards achieving interoperability, simple integration, internetwork compatibility, and seamless networking; this simple assertion is critically important as local area networks become switched infrastructures supporting peer-to-peer workflow solutions, client/server revolutions, and enterprise integration of information processing and Internet distribution. While standards promote reliability, product growth, compatibility, and interconnectivity for local area networking, they are by far of greater importance for interconnectivity of heterogeneous enterprise networks. Standards are absolutely necessary for development of effective network management tools for Fast Ethernet variants and future network transmission protocols.

Reading Network Diagrams

This is a quick lesson on reading network diagrams. Thick lines indicate a network backbone to connect all network devices. Network devices, or nodes, include PCs, workstations, printer ports, servers, and all intermediate network devices (such as wiring hubs, repeaters, bridges, routers, switches, and gateways) are indicated by boxes. Cabling between a device and a backbone or hub are indicated by thin lines. Thin lines could represent either 10Base2 "T" connections, 10Base5 transceiver cables, or the typical twisted-pair wiring commonly used now.

A network diagram can be an exact physical representation showing component placement and actual paths for the wiring. This is called a "physical" diagram and is usually a layered floor plan. Each layer represents a different function, from basic structure designs to HVAC ductwork, water and sewer lines, power lines, fixed partitioning, movable partitions, and data wiring. You will see floor plans throughout this book. Figure 2.3 illustrates a typical modern office building. This same octagonal floorplan is shown at various sizes to give a scaling perspective to other diagrams. Such a diagram might be imported to Windows from AutoCAD as a bitmap.

Sometimes a blueprint shows too much. Even when layers are extracted to highlight key points, a blueprint can still be too big. The network design is therefore condensed to its salient points, which are typically network devices, the wire between them, and the logical way in which these devices are attached together. This stick-figure diagram is called a *logical* representation. It is the condensed diagram for the network. Issues of scale, color, placement, and measurement are unimportant to the diagram; the logical relationship between devices is the point. The logical

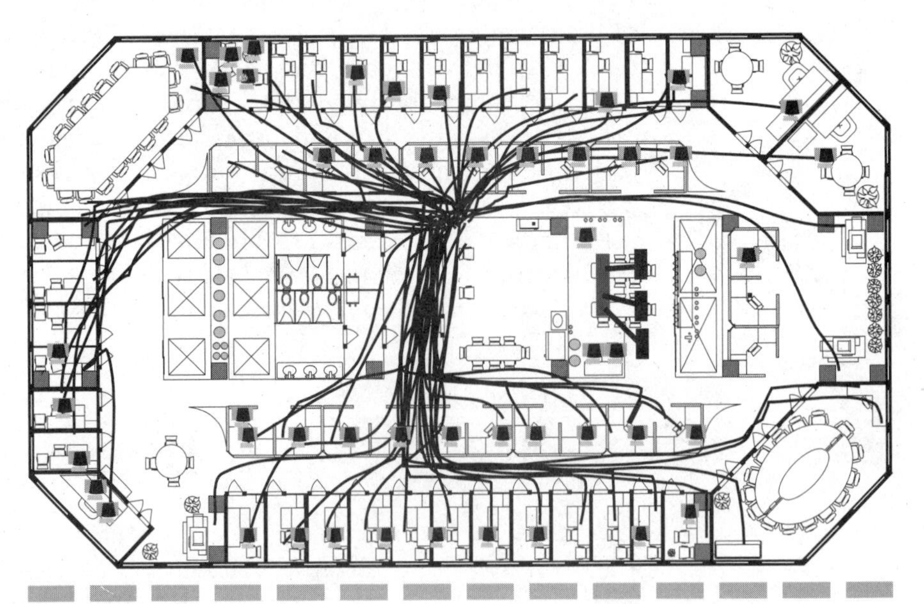

Figure 2.3.
Physical network diagram.

diagram shows how network nodes connect to one another and the relative placement of intermediate nodes such as hubs, repeaters, bridges, switches, and routers. You will see logical designs throughout this book, because they convert complex relationships into stick figures. They are cartoons for all intents, as shown by Figure 2.4.

10Base Historical Standards

The original 1975 Ethernet standards are now referred to as "Ethernet 1.0" to distinguish from the "Ethernet 2.0" agreement made in 1980 by Xerox, DEC, and Intel. This corporate partnership has since become codified by the Institute for Electrical and Electronic Engineers (IEEE) as a transmission protocol by the 802 protocol committee. This standard is represented by these actual definitions; IEEE 802.2 (media access layer), IEEE 802.3 (topology), IEEE 802.3u (Fast Ethernet), IEEE 802.9a (Isochronous Ethernet or IsoEthernet), IEEE 802.11 (wireless LANs), and IEEE 802.12 (100BaseVG-AnyLAN).

Ethernet transcends just the IEEE as it has been ratified and supplemented by ECMA-80 (General Assembly of European Computer Manufacturers Association), ECMA-81, ECMA-82, ECMA-72, and the ISO OSI communications reference model. EIA/TIA (Electronics Industries Associations/Telecommunications Industries Association) 568 and EIA/TIA 569, and EIA/TIA technical service bulletins (TSB) addendums 39, 40, and 67 provide effective standards for premise wiring with twisted-pair and fiber with Fast Ethernet. Forthcoming proposals also include testing protocols. The early Ethernet corporate partnership was prescient of current and future needs for conformity and interoperability.

Ethernet comprises layers 1 and 2 of the Open Systems Interconnect (OSI) model. Refer to my book *The Ethernet Management Guide* for additional background on the OSI frameworks. Layers 3 through 7 are not part of Ethernet, although Ethernet is often packaged with the TCP/IP protocol; TCP/IP provides network-layer functions. The XNS, DECnet, LattisNet, HPnet, and other vendor-supplied protocols also communicate with Ethernet. TCP/IP was commissioned by the Department of Defense (DoD) for the ARPAnet and later implemented on the Internet network. Although standardized by the IETF (formerly the IAB), it also is supplied by a rapidly growing list of manufacturers. In the future, you will likely see connectionless UDP supplant TCP/IP for remote and wireless transmissions, because it is less verbose and thus requires less overhead.

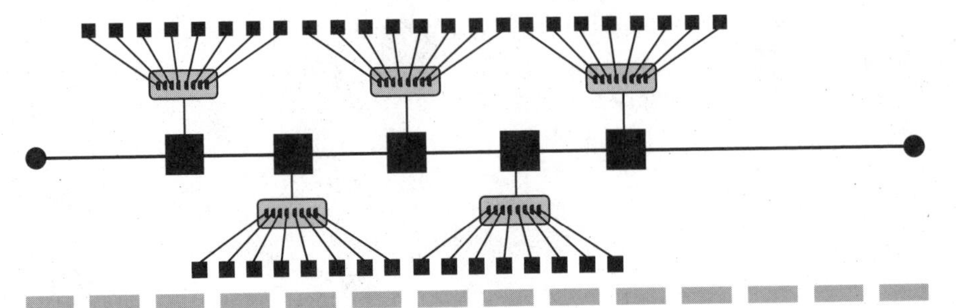

Figure 2.4.
Logical network diagram.

The great virtue of TCP/IP is its simple structure and its ready implementation within the memory and speed limits of most computer platforms. TCP/IP is much less verbose than NetWare IPX/SPX, and has another great virtue of minimizing hop traffic overhead between the subnets when migrating from Ethernet to Fast Ethernet. The great disadvantage of TCP/IP as implemented is that it does not conform rigorously to the OSI stack protocol—the layers, as previously mentioned—with the result that the ISO OSI model is falling into disfavor, and that the IPv4 does not support a large enough unique address space for the 100 million current network devices or the forthcoming flood of new ones. IPv6, better known as IPng, is slowly replacing IPv4 to support billions of unique and (registered addresses). Although TCP/IP does support a *sliding window flow control* so that more data frames can be sent before a confirmation is required, it still requires a lot of overhead.

Ethernet is represented only by hardware, and it is fortunate that the hardware is very malleable and in fact separable from the media. The protocols build on the basic hardware. The operating system components at each node that provide the communications functions build on the Ethernet controller (transceiver and bus controller or integrated NIC) function. Ethernet, like a telephone, will handle different application software; and just as with any phone that accesses a network of other phones, it must adhere to certain formats.

Fast Ethernet Physical Formats

This section describes the Ethernet hardware, the variations, and the physical differences among them. Although Ethernet is strictly a hardware protocol, the physical hardware is really divorced from the logical implementations. You emphatically need to understand the Ethernet and Fast Ethernet variations and how they interrelate because cost, performance, maintenance, and reliability factors mean that one variant or architecture will better match your needs.

The logical mechanical standards for Ethernet include baseband coaxial cable, twisted-pair, radio frequency and infrared, broadband coaxial cable, and optical fiber. The original Ethernet was a bus-based design now obscured by star-based premise wiring standards. Although the original frame sizes and formats persist, the only identity remaining of the original Ethernet is the collision-detection protocol and timing limitations defined by the structure and collision-detection protocol. This vestige is almost hidden with the advent of duplexed- and switched-Ethernet variants. However, the timing parameters still define the physical, logical, and distance ranges for 100 Mbps + 1Gbps variants.

Because 100-Mbps Ethernet is precisely the same in every regard as 10-Mbps Ethernet except that all times are one-tenth as long, you can have no more than 5.12 ms as the worst-case end-to-end propagation delay. This does not mean your new Fast Ethernet will always deliver data within that timing limitation. Switching, routing, and collisions add to actual data delivery times, although Fast Ethernet is likely to provide performance that is better than standard Ethernet for stand-alone LANs.

Internal electronics in hubs and wiring delays limit hub delays to 2 ms. This has a profound effect on the viability in the near term of true node-switching hubs at 100 Mbps. While switching is useful for increasing bandwidth, there is still a serious delay at the central switch. Fast Ethernet switching at 100 Mbps looks like a desirable method to increase bandwidth in LANs when switching times are less than 10 ms. However, switches based on new *Application-Specific Integrated Circuits* (ASIC) technology generally provide switching times that are about 30 to 60 ms, while older switching technology provides 200- to 300-ms switching times, equivalent to routing delays. However, 10- to 20-ms switch delays are an order of magnitude larger than average 100Base-T delays. This is still a lot faster than the 40- and 80-ms access delays seen in Ethernet or Token-Ring; migration to Fast Ethernet should resolve many network bottlenecks. Because of these significant delays, you need to understand the ramifications when you plan any migration from ARCNET, Ethernet, or Token-Ring to any Fast Ethernet variant; you could mistakenly trade more bandwidth for longer delays, and thus lower overall performance. The reality is that Fast Ethernet switching is faster than the 10-Mbps options, but not as efficient or fast as 100Base-T alone or FDDI.

Thus it is useful for trading bandwidth where you need it at the expense of longer latencies.

These bandwidth and delay limitations are determined by the clock rates, the speed of electrical transmissions, and cable lengths, although throughput is limited by collisions and packet overhead. Networks built with coaxial cable must be terminated in its characteristic impedance at each end. Even networks built with hubs and twisted-pair require proper termination, although this is usually automatic. You will want to watch how deep you cascade hubs, as you do not want to create an architecture that exceeds the collision window.

CSMA/CD protocols, particularly the window during which a valid collision can occur, place certain limitations on the physical channel in terms of length and complexity. These limitations specify maximum signal propagation times, and as a consequence, maximum cable lengths, because cable lengths and propagation time affect the slot time as defined in the data link layer of the OSI model. Overall, you can have no more than 51.20 ms as the worst-case end-to-end propagation delay on any 10-Mbps Ethernet network. 51.20 ms loosely corresponds to the worst-case signal transmission time end-to-end over three 500-m backbones interconnected by two repeaters. This time includes all the travel paths, including:

- Encoding/decoding packets
- Transceiver receive and transmit paths
- The transceiver collision path
- Coaxial cable (backbone, backplane, and hub electronics)
- Repeater electronics
- Repeater cabling
- Carrier sense lag
- Collision detection lag
- Signal rise times

Although these times were based on 10Base5 physical-layer components, the same limitations are in force for 10Base-T; the collision window is the same. Basically, you can have no more than 20-ms delays through a hub and another 12 ms in repeater or backplane delays. The 20-ms allowable delay in a hub is the speed yielded by the fastest 10-Mbps Ethernet switches when interconnected to a standard backbone architecture. Some slow switches cannot perform within the allowable limits of Ethernet, and create late collisions on shared media segments. Because Fast Ethernet is just this, but with a clock that is 10 times faster, all limitations for Fast Ethernet scale as multiples of 10. However, because signal propagation times are fixed by the speed of electricity moving down a wire, wiring lengths do not change very much.

It is useful to review standard Ethernet limitations. 10Base5 or 10Base2 backbone cable can consist of many spliced *sections*, the smallest unit of cable. One or many sections constitutes a cable *segment*, which serves a stand-alone network, subnet, or subnetwork. The segment may not exceed 500 m between the endmost transceivers for 10Base5, or 185 m (nominally 2×100) for 10Base2. The shorter length for 10Base2 is not a function of slower transmission speed, but a factor of the lesser cable shielding. By the way, you could have a backbone for 10Base-T, but due to the collision window parameters, any such backbone would be limited to 5 m in length.

This also limits the wiring paths inside a 100Base-T hub to the same 5 m or corresponding signal propagation through the circuitry. Figure 2.5 illustrates these limitations. Many sections can be spliced to construct an Ethernet segment. Sections, in fact, ease the burden of debugging most types of network hardware faults. Note the proper termination required to complete a segment.

There may be a maximum of two repeaters in the path between any two stations. Repeaters do not have to be located at the ends of a segment. Repeaters can be used to extend the length of the segment or to extend the one-dimensional bus topology into three-dimensional configurations encompassing different groups, floors, and buildings. Hubs and switches occupy transceiver positions on each cable, and count towards the maximum number of transceivers on a segment (100 units). The total worst-case round-trip delay for all coax in the Ethernet system is 13 ms. This corresponds to 1.3 ms for Fast Ethernet.

Although this expresses a continuous two-dimensional backbone, it is also technically possible to create a structure with an unlimited number

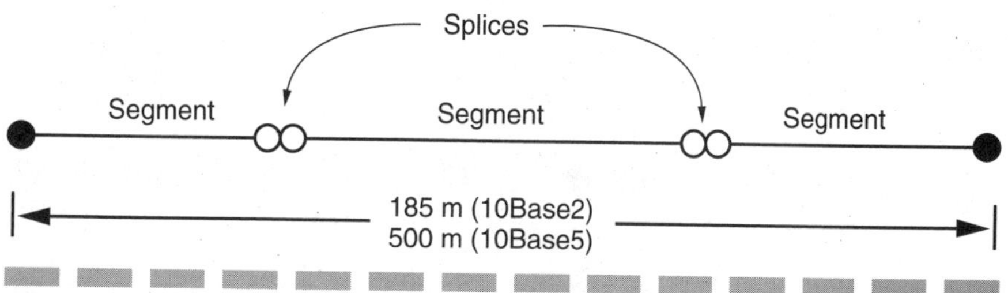

Figure 2.5.
Maximum 10Base2 and 10Base5 Ethernet segment lengths.

of repeaters, as Figure 2.6 shows. This architecture underlies all 100Base-T, 100Base-T4, and 100BaseVG wiring hubs and the wiring hub chassis, as well as all switched architectures. The repeater function might not be explicitly defined, but it is a function of the attachment of each wiring hub adapter into the hub backplane, stackable hubs, or enterprise backplane switches. The repeater electronics can be built into the hub adapter itself.

The propagation velocity for a signal on twisted-pair is only 58 percent of light speed. As such, 100Base-T and 100Base-T4 on unshielded twisted-pair supports a maximum of 100 meters between the wiring hub and each node. This is formally represented as 90 m for the lobe cable, 3 m for the connection from the wall plate to the NIC, and 6 m for wiring closet jumpers. Figure 2.7 illustrates these lengths. Figure 2.8 shows these lengths for 100BaseVG. Superb hub and NAU units provide strong and well-differentiated signals, while quality wiring will carry the Fast Ethernet signal within acceptable distortion, capacitance, NEXT, and crosstalk limits. While it is possible in practice to extend the lobe lengths beyond 90 m, extended configurations are likely to increase the collision window or extend other latencies beyond what is acceptable on your network.

It is important to realize that although 100BaseVG and 100Base-T4 use four pairs for transmission, the wiring length limitations are different. 100Base-T4 is limited to 100 m, just like 100Base-T. 100BaseVG supports 200 m. While 100Base-T4 provides some opportunity for migration from

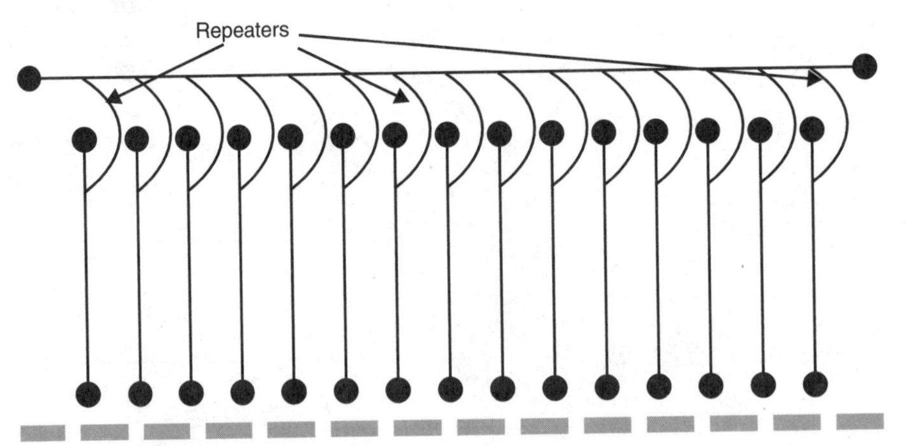

Figure 2.6.
No more than two hubs between any nodes; but many hubs can be used within that limitation to create complex LAN topologies.

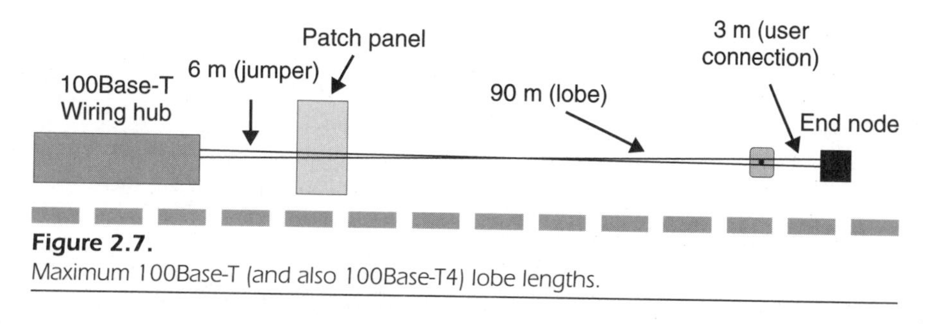

Figure 2.7.
Maximum 100Base-T (and also 100Base-T4) lobe lengths.

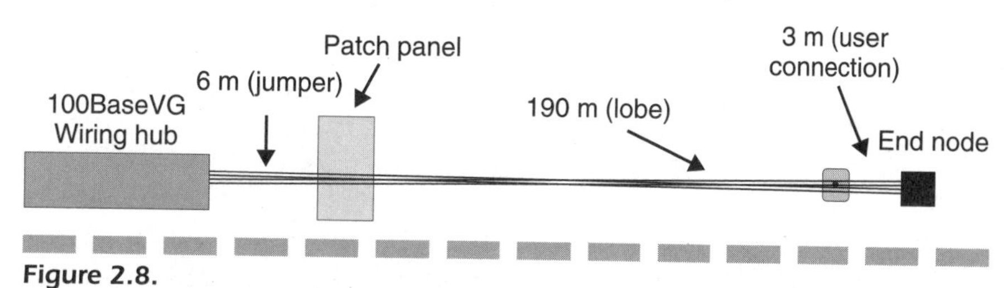

Figure 2.8.
Maximum 100BaseVG hub to node lengths. Note that 100BaseVG doubles standard 10Base-T and 100Base-T lobe lengths at the expense of requiring extra pairs.

10Base-T without rewiring, it does not provide greater wiring spans. HP designed 100BaseVG to appeal as a migration technology from STP-wired Token-Ring (in addition to 10Base-T), so the extra spanning coverage fits existing configurations. You still need four pairs for 100BaseVG.

There may be a maximum of 1000 meters of point-to-point coaxial link anywhere in the system supported by half-repeaters, as illustrated in Figure 2.9. This will typically be used to link separate segments on different floors in distant buildings. You may not place any end-user nodes between half-repeaters; this is strictly a point-to-point connection. The common application for this technology is to connect workgroup hubs to the enterprise hubs. Although the vendor is unlikely to call the connection a "half-repeater," any coaxial connections between hubs are bound by this specific limitation. The point-to-point propagation velocity is assumed to be 65 percent of light speed in the worst case.

The maximum optical fiber point-to-point coaxial link (FOIRL) may be 2000 meters, as illustrated in Figure 2.10. This is a function of the fiber-optic components and standardization with other fiber technology. The FOIRL connection, an ISO level-1 half-repeater, can interconnect 10Base2, 10Base5, 10Base-T, 100Base-T, 100Base-T4, and 100BaseVG. This is a useful technology for creating a multifloor backbone or a horizontal

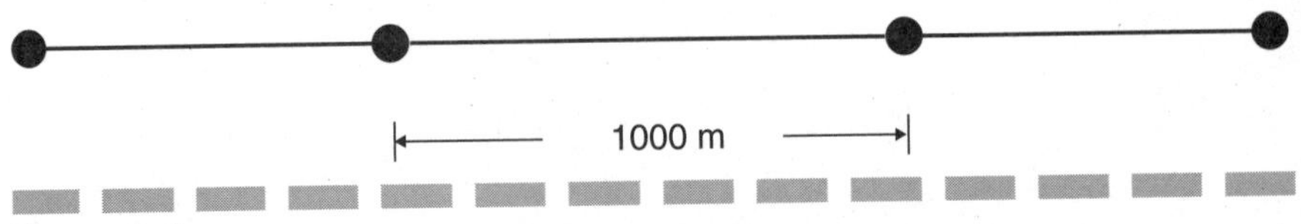

Figure 2.9.
Maximum Ethernet half-repeater point-to-point coaxial link.

campus backbone, because of the distances spanned by optical fiber. While it is not as flexible as a true FDDI backbone, it is simpler in that many vendors (3Com, Cabletron, and Bay Networks to name but a few) sell FOIRL repeaters and optical fiber 10Base-FX and 100Base-FX Ethernet transceivers. When planning your migration, make sure the vendor can provide 100-Mbps linkages, not just 10-Mbps ties, as the slower connection could create significant bandwidth and performance hits.

100Base-T and 100BaseVG hub cascades can generally be cascaded from two to five units deep, but no more. The number cascaded is a function of the hub design and electronics. Because Ethernet and Fast Ethernet specify no more than two FOIRL links between any two subnetworks, split the electronics into four, and this provides the five layers (when the top layer is included). Because the Fast Ethernet limitations are really defined in terms of maximum delays, vendors can work within these parameters to create more flexible architectures. The Ethernet, and hence Fast Ethernet, timing parameters are the functional product design limits rather than stated wiring length or cascaded architecture depth standards.

A cascade means that a lobe connection on a hub is attached to a port on yet another hub, as Figure 2.11 illustrates. As a result of this limiting architecture—and you are limited by it—many so-called stackable hubs or chassis hubs provide a separate connection to the master backplane.

The clue for backbone support is a separate attachment unit interface (AUI) connector for 10Base5 or a BNC connector for 10Base2. If a hub is termed a "stackable" or "enterprise" hub, it usually has a physical connection to simplify integration into the internetwork. However, some products provide this outlet merely for integration or interconnection with existing Ethernet, and to simplify stepwise migration. You need to check the bandwidth for this backbone, because 10Base5 and 10Base2 limit bandwidth to 10 Mbps. FDDI or FOIRL dual fiber connectors support 100 Mbps rather than 10 Mbps on products designed specifically for Fast Ethernet enterprise integration.

In many cases with chassis-based hubs, this backplane has more usable bandwidth than Fast Ethernet or Switched 10 Ethernet, at times up to 10 Gbps. Most chassis-based hubs provide from 60 to 240 Mbps in usable bandwidth depending on how it is partitioned. The master backplane can support as many hubs or hub adapters as you realistically want to add to it. Although it is possible to construct a single 100Base-T Ethernet with several hundred nodes, performance will likely not be very good at all. If you can microsegment that network into subnets with router or bridging modules that stack with the stackable hubs or fit within the chassis hub, you are likely to see better performance and be better able to tune configurations to balance loads.

The point about cascading 100Base-T hubs is that you do not want to support that cascaded architecture, as it provides an inferior design. All the backbone traffic gets concentrated at the top of the cascade, often overloading the 100-Mbps bandwidth of segment connectors or the larger bandwidth on the hub backplane. Instead, you want to attach the hub modules with support for server and user nodes directly to a single backbone. Indeed, this backbone is equally serviceable as either a

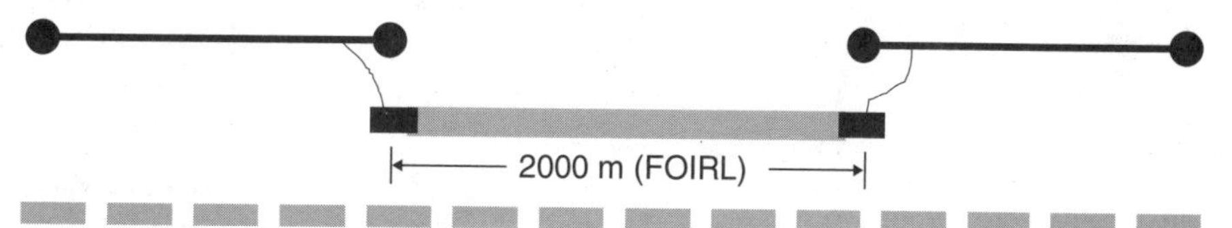

Figure 2.10.
Maximum Ethernet point-to-point FOIRL link.

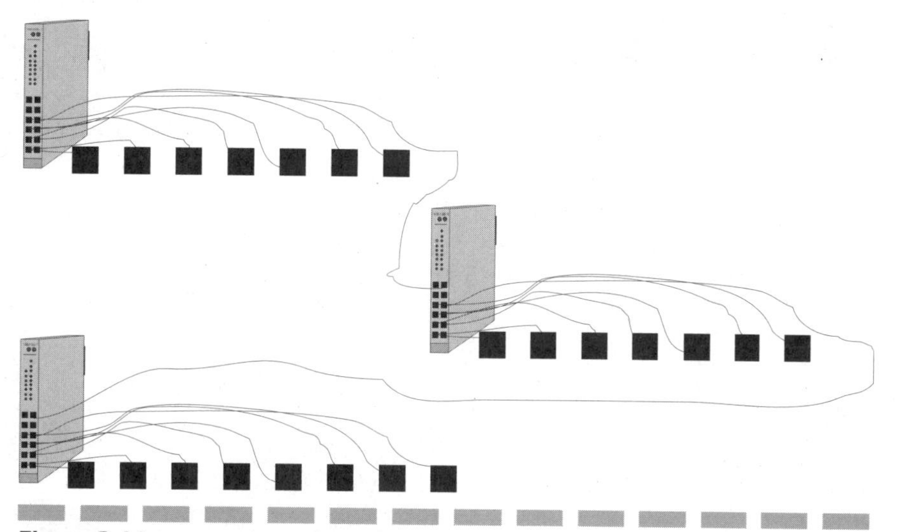

Figure 2.11.
A cascade of 10/100Base-T wiring hubs.

backplane chassis or stackable hubs with a separate attachment interface to its own backplane. Note that stackable hubs are nothing more than a vendor packaging issue, and stackable hubs perform no better and no worse than wiring hubs or enterprise switches.

The big benefits to stackable hubs is in the ease with which larger networks are interconnected when the stackable hubs have a separate backplane to interconnect multiple hubs. Another issue with stackable hubs is that they distribute the points of failure. This can either increase or decrease overall network reliability. If you opt for stackable hubs for simpler distribution and increased design flexibility, I suggest you confirm support at least for SNMP or your network management software, and optionally for SMNPv2 to simplify network management. Many 100Base hubs do not provide built-in network management or interoperate well with Cabletron Spectrum, HP OpenView, Sun NetView, or products from Tivoli (IBM's wholely-owned acquisition). You need basic network traffic statistics, as described in Chapter 5, to balance loads, track reliability, or wisely restructure the network. Do not back yourself into a corner and assume that migration to Fast Ethernet will be better and will also remain better.

Backplane performance will surpass that of cascaded hubs by lowering the collision rate. In general, the collision rate is a function of overall LAN latency and traffic levels. In extreme designs, as with cascaded hubs, you merely increase the length of the collision window. The alternative to the cascade is a backbone architecture, as illustrated in Figure 2.12. The fundamental benefit to this design is that you can easily migrate 10Base-T hubs to 100Base-T, 100BaseVG, switched 10, switched 100, or FDDI over copper, or replace the Ethernet and Fast Ethernet with local ATM when that technology is available.

Nevertheless, a cascaded design is not too different from a backbone connecting to the wiring hubs, because both designs create a single LAN on one logical channel. When the shared media bandwidth is exhausted by the density of nodes on the backbone, it will not matter what that backbone architecture is. The only solution is microsegmentation or a uniformly faster infrastructure with a greater bandwidth.

Basic LAN Architecture

The next sequence of illustrations shows the original bus Ethernet design and how it has been extended with 10Base-T and chassis hubs.

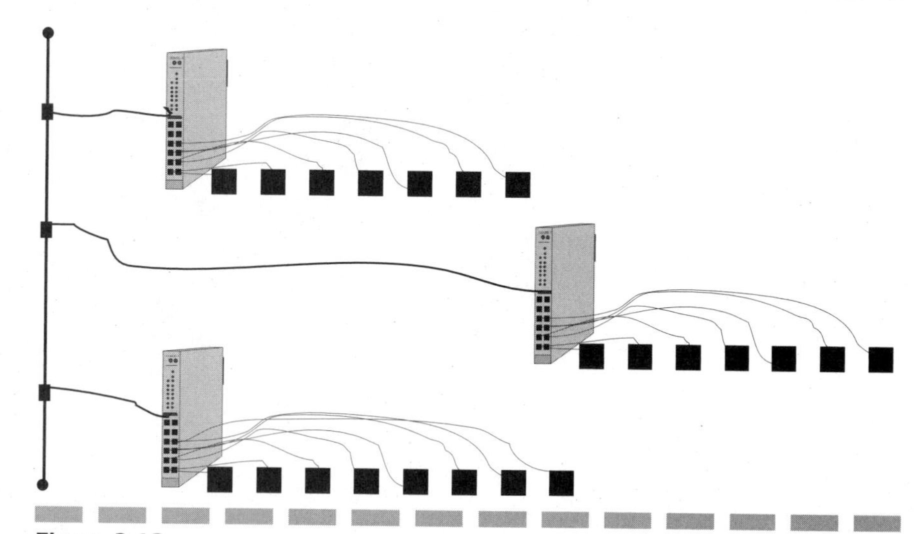

Figure 2.12.
Create a backbone architecture rather than cascade hubs. The backbone can be a "slow" Ethernet coax (as shown), an FDDI ring, 100Base-FX, or an enterprise hub.

The purpose of these illustrations is to show that some understanding of fundamental Ethernet design is required for successful network installation and management. While it might not be important to know the signal-to-rise time limits for signal propagation, it is very important to understand the abstract building-block components for Ethernet LANs. Figure 2.13 shows the fundamental Ethernet as the bus.

Even 10Base-T and 100Base-T retain the logical bus design. (Note that the wiring is star-shaped, with all wire directed into the hub from end-user nodes.) The hub contains self-terminating switches or cutout switches for unused ports. Although the 10Base-T signal is very different from the 10Base2 or 10Base5 signal, the AUI port and 10Base-T or 100Base-T electronics creates a backbone bus nonetheless, as illustrated by Figure 2.14. You can envision 100BaseVG with the same logical design, although this is not a completely valid comparison because, in 100BaseVG, the demand priority protocol controls access to the common bus.

Stackable hubs and chassis-based hubs retain a bus backbone architecture. The stackable hubs create this backbone with AUI connections, a special channel bank wire, or an additional snap-in module that ties the hubs together. The chassis hub literally has a built-in backbone, typically referred to as the *backplane*. The cards, which fit into the chassis, have repeater electronics that link all nodes on the local segment into the same LAN backbone. It is important to realize that these chassis-based

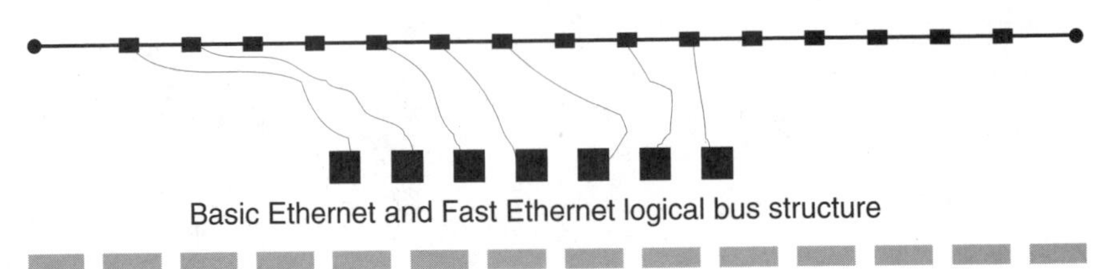

Figure 2.13.
Basic Ethernet LAN on bus architecture.

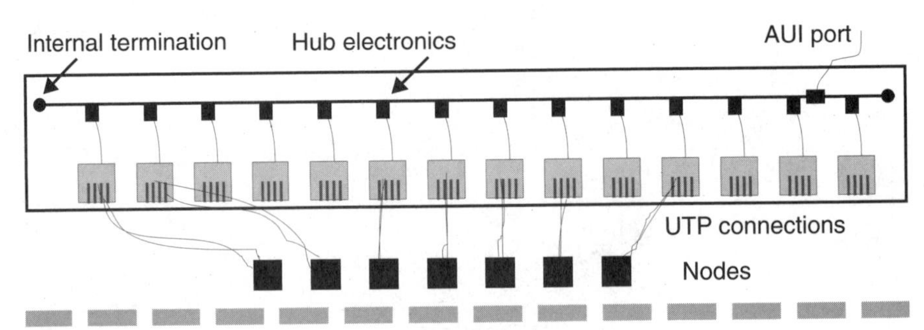

Figure 2.14.
Basic hub-based LAN retains bus architecture internal to hub box.

or stackable hub architectures do not provide microsegmentation unless it is explicitly enabled. In other words, all nodes in Figure 2.15 are still on the same LAN, which is also a backbone bus.

Figure 2.16 shows a chassis-based hub with a high-speed segmented backplane for multiple-channel support. The hub electronics include wiring hub support as with any other hub, but this unit also includes some sort of router hardware and software to partition the network into subnets. Configuration in hardware and/or in software will create multiple LANs, each on its own microsegmented backbone bus. While it is possible to create a single backbone on a bus supporting hundreds of nodes even with 100Base-T and 100BaseVG, this is likely to yield poor network performance.

Performance can be improved with switching by microsegmenting traffic over a segmented backbone. At least three different fundamental types of switching occur. Performance varies with each type, and is also a function of network load, speeds, and other design and configuration issues. First, you can have a manually configured microsegmentation, and this is really no different from partitioning the LAN into static subnets. Second, you can "switch" traffic onto different backbones by establishing an electronic preconfiguration of nodes on the different backplane channels. Third, the system can switch nodes through a routing mechanism (cut-through or store-and-forward) based on various routing algorithms. Nonetheless, these designs of hub-based switches still retain the original Ethernet bus backbone, as shown in Figure 2.17.

Even when switching is pushed to the limit of possibility so that two nodes and only two nodes exist on a private LAN just long enough for a complete transmission, the switching architecture creates a virtual circuit that still looks like a bus-based LAN, as shown in Figure 2.18.

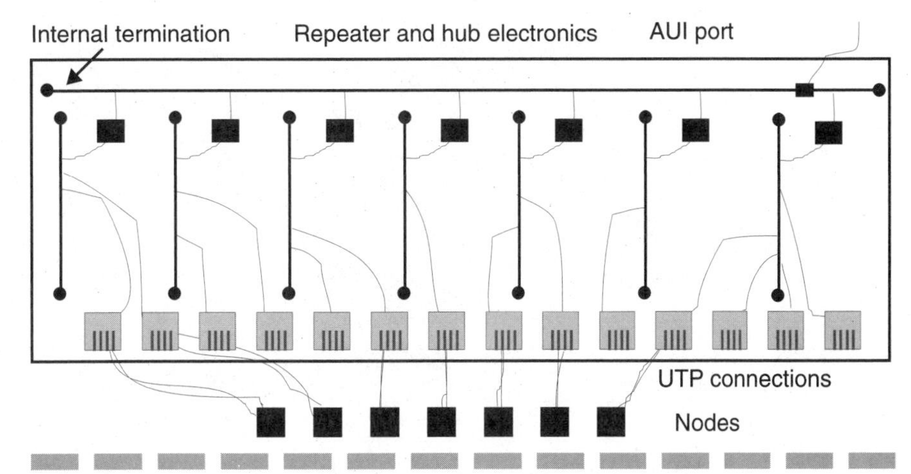

Figure 2.15.
Hub-based LAN extends bus architecture with internal repeaters.

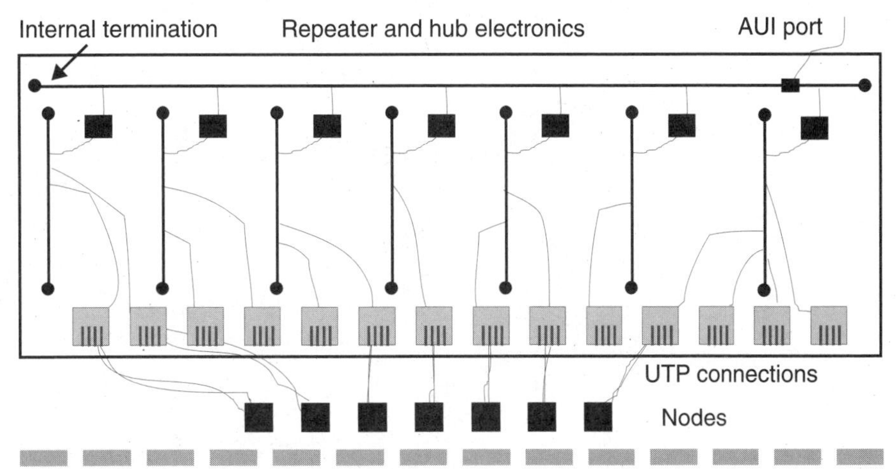

Figure 2.16.
Hub-based LAN extends bus architecture with internal routers.

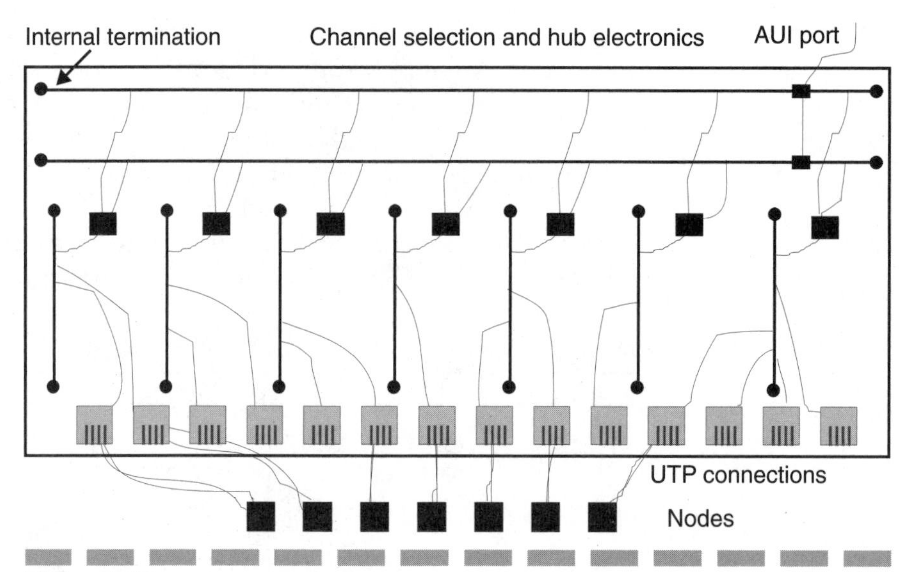

Figure 2.17.
Hub-based LAN extends bus architecture with internal switching.

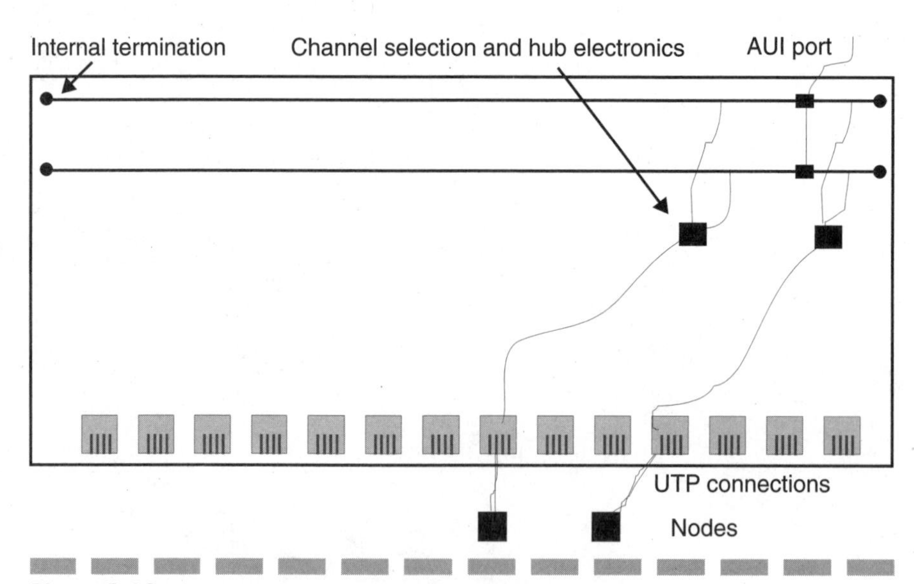

Figure 2.18.
Hub-based node switching creates a virtual LAN for two nodes.

That node-switched design looks like a series of matrix bridges overlaid on a standard Ethernet bus backbone, as illustrated in Figure 2.19.

Virtual LANs, or VLANs, are not a Fast Ethernet technology. Although a few Fast Ethernet VLAN products are making it to the market, you need to differentiate the VLAN component from the Fast Ethernet component. VLANs promise simplified network management from a wiring closet or network management station. In reality, the technology is not yet mature because effective remote management and network management is not currently available. It will be soon, because the VLAN technology is needed for network management, performance optimization, and workload reductions. It is also a step on the road to ATM networking. VLANs do not improve network performance, increase bandwidth (except by physical microsegmentation), or make the network more realiable; instead, this technology simplifies and amplifies network management, an important issue for large networks.

One of the most time-consuming tasks for network management is handling the moves, adds, and changes with users, user stations, and wiring. Virtual connectivity separates some part of the physical wiring tasks and makes them logical. When a user station is moved, added, or somehow changed, imagine how easy it can be simply to flick a switch to retain the same network workgroup connection, assign a new network

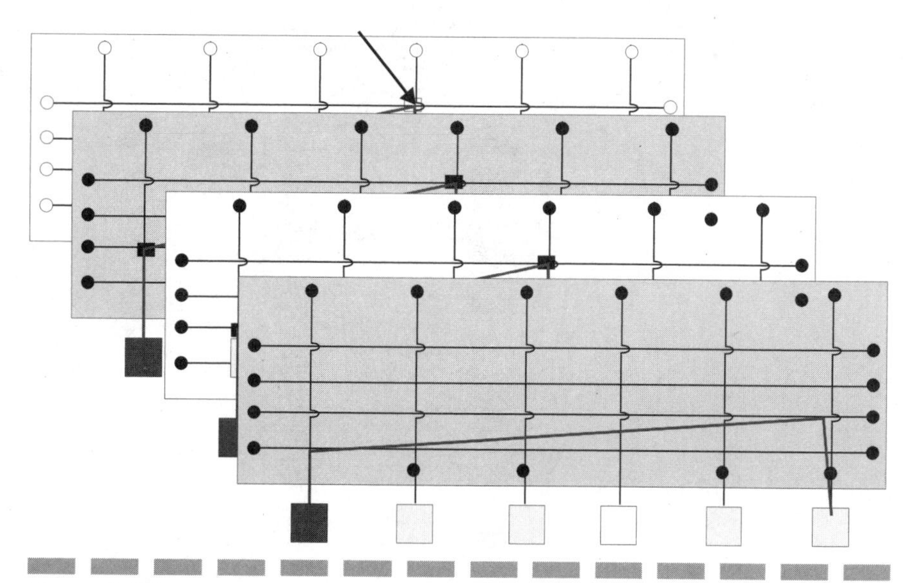

Figure 2.19.
Virtual LANs are composites of a matrix array of bridges allied with a bus-based backbone.

address, and balance loads. The reality is that virtual connectivity creates a layer of abstraction between the physical wiring and hubs and the logical network design.

Think of VLANs as the old-fashioned telephone switchboard and the operator plugging retractable wires into jacks to connect calls. It is a big panel of possible connections. Every line in gets a jack, and every network device gets a retractable cord. Because lines are nodes, you have a one-for-one relationship. Figure 2.20 illustrates this concept.

You might note that it generally makes no sense to plug a node cord into its own jack. However, this is called the "loop-back" test, and it is useful to confirm functionality or isolate a failed node from the network. The loop-back test is shown as the diagonal line. As you see, the "switch" is not really one switch at all, but rather n^2 switches where n is the number of ports. Hence, a two-port switch has four switches, and a 24-port switch has 576 switches. These massive switching grids become arrays. Because a mechanical cord and person is slow and not always perfect—that is the problem with standard 10Base-T, 100Base-T, or 100BaseVG wiring hubs—you want to automate the switching process. Relays are too slow, tend to require too much power, and do not retain state, so the switch array is built into a chip. The software matches physi-

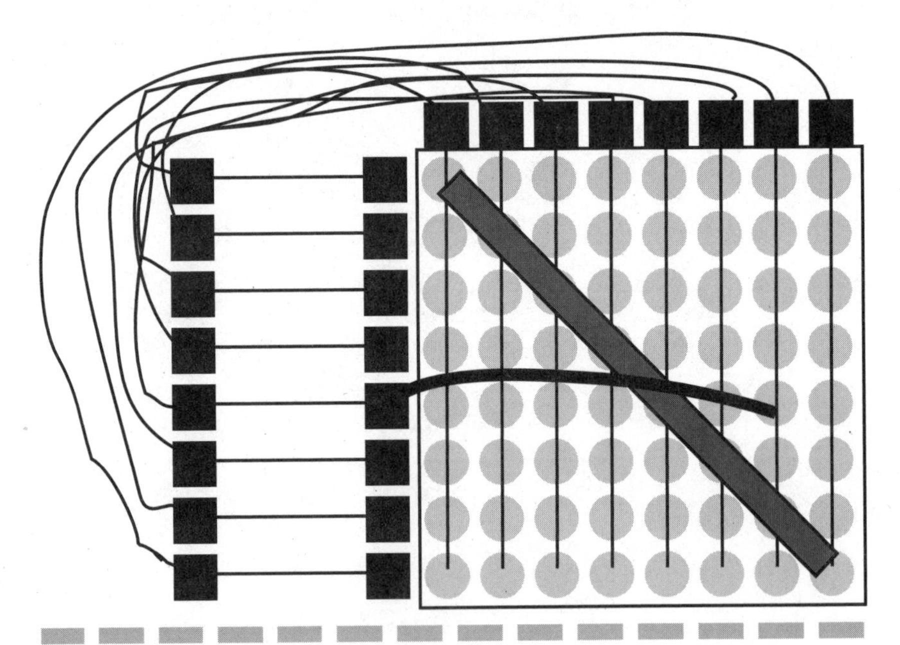

Figure 2.20.
The VLAN physical layout with retractable cords and jacks.

cal port assignments with a secondary port designation. This might also be associated with a table of network ID addresses, either an IPX or IP address. The next table shows the layers of abstraction.

Physical port	Port usage designation
1	142.67.220
2	142.67.113
3	142.64.150
4	142.64.005
5	142.64.008
6	142.65.105
7	142.67.006
...	142.64.018
n	142.64.055

Network addresses should be unique for LAN-to-LAN routing. If you have NICs with duplicate IP addresses (this is extremely rare), replace the NIC. If you assign duplicate NetWare addresses because you cloned a server and all its client setups, you will need to correct that. Because most LANs start out small and use the default server numbers and addresses—a new administrator rarely knows or cares about these seemingly insignificant nuances—you might well need to renumber and rationalize the node addresses. If you look at the addresses carefully, you will note three different LAN subaddresses.

This means the VLAN is supporting at least three (there might be more) LANs. It is also possible that some of the nodes think they are on different networks and create a routing hop at the server. I have seen this performance problem in a few networks. It is usually invisible unless you decode packets with a protocol analyzer and map the addresses to physical nodes. The extra hop adds latency to message delivery and adds more work to the server. In any event, this VLAN architecture provides no performance boost over a comparable hard-wired subdivision in a wiring hub. You could also provide the same microsegmentation by jumpering the lobe wires to different wiring hubs. If you are using a Cabletron MMAC+, Synoptics 8000, or similar multislotted backplane, you would assign each wiring hub on a card to a different channel. The microsegmentation is provided only by patching, as illustrated in Figure 2.21, and you can achieve the same physical performance effects if not the managerial ones.

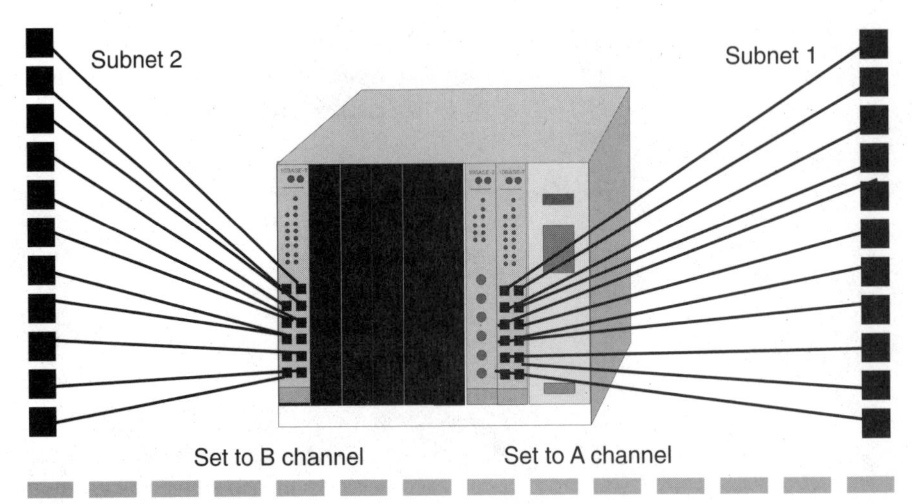

Figure 2.21.
Creating a "VLAN" without the equipment.

The important reason for virtual LANs is that it lets network management group users together in a logical rather than physical manner. Users sharing the same server, printers, and other common network resources can be hooked together on one workgroup LAN. This bypasses the bandwidth blues so common when physical wiring creates network limitations. When multiple VLANs connect together, there is a need for standard identification for packets going to or coming from each VLAN segment. One proposal based on the IEEE 802.10 frame tagging standard adds extra bits inside the data portion of the packet for routing information. This will likely slow routing and switching, as this information must be decoded at every device. A second option synchronizes MAC address with a database, a method not unlike a routing table. However, there are three commercial switching methods to define a VLAN, port, MAC, and IP; standardization is lacking.

A duplexed LAN creates two bus backbones, one for transmission and the other for reception. When virtual switching is added to duplexed LANs, you have dedicated bandwidth so that two nodes and only two nodes exist on a private LAN just long enough for a complete transmission with twice the shared-media bandwidth, as Figure 2.22 illustrates.

Duplexed connectivity increases usable bandwidth for hosts and network servers. It is sometimes useful for engineering workstations and user PCs. If you are considering the transition from a bottlenecked Ethernet to Fast Ethernet, it can be more cost-effective to increase the bandwidth for just the servers with duplexed connections and intelligent microsegmentation, as most users do not need duplexed, switched, or 100-

Mbps bandwidth. When hosts and servers talk to each other, dedicated duplexed 10-Mbps or 100-Mbps connections between just these components might provide sufficient bandwidth to relieve the traffic bottlenecks. A number of transitional products mix standard Ethernet with duplexing and switching. Kalpana, bought by Bay Networks, the leader in duplexed hubs, provides several standalone and stackable hubs with 10 standard Ethernet ports and two duplexed or switched ports. Even some older products, such as the Cabletron MMAC or MMAC+ chassis, give you the option to mix Token-Ring or Ethernet with these Fast Ethernet solutions.

Although there is some increase to basic bus bandwidth as the figures advance in this series, it is also important to recognize that each design increases the latency for signal transmission and the window for Ethernet collisions. Although some switching and all duplex technology disable the collision detection mechanism, signal latency for complex and mixed-speed networks is an important concern.

The Wiring Hub

Although the *wiring hub* is necessary to provide base-level service for a 10Base-T and 100Base-T network, the hub also serves as an effective

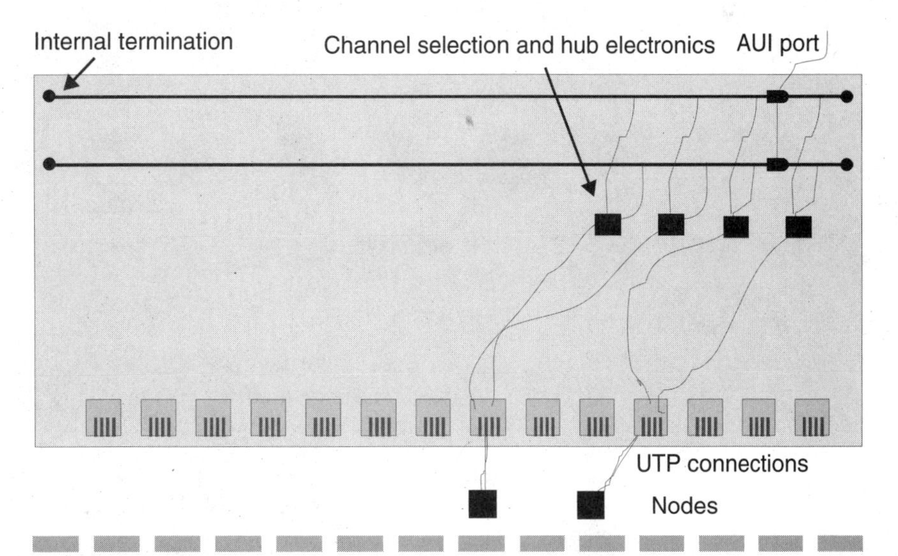

Figure 2.22.
Duplexed technology uses two rather than one bus backbone even when that bus backbone is enabled through internal electronics.

expansion unit to a preexisting 10Base2 or 10Base5 network. The wiring hub is essentially a multiport Ethernet repeater. A single transceiver and drop cable connecting a hub services as many ports as are in that unit without the need for individual transceivers. Twisted-pair jumpers and lobe wires connect individual node workstations to the hub, or a micro-MAU and AUI drop cable converter bridge the media differences. While this does create a mixed-media configuration, it is acceptable and provides a realistic transition to twisted-pair premise wiring. It provides an acceptable and realistic means to add 10/100Base-T nodes to a preexisting 10Base5 or 10Base2 network. Figure 2.23 illustrates a typical twisted-pair wiring hub.

This represents the simplest installation for integrating any Fast Ethernet twisted-pair network configuration into a preexisting Ethernet network. It is more involved to interconnect Token-Ring with Fast Ethernet because of both packet and frame size and format differences. The generic solution is to install a bridge to interconnect the new with the old, although NetWare, Windows for Work Groups, and Windows NT support ODI and NDIS to simplify communication standardization

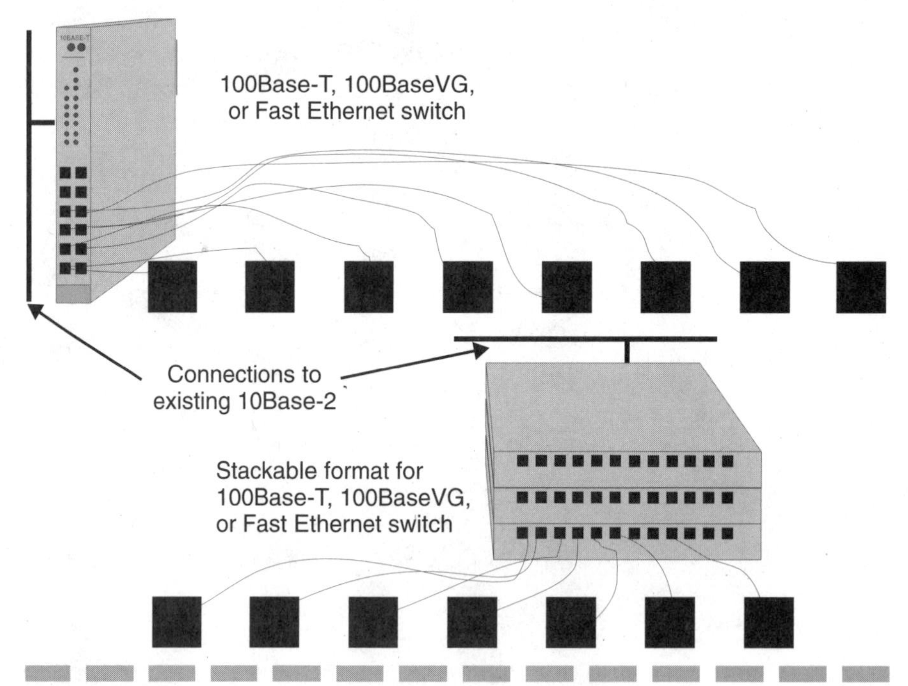

Figure 2.23.
A typical twisted-pair wiring hub.

(despite the lack of network connectivity standardization). A simple Fast Ethernet network requires no coax cable, transceivers, connectors, or terminators. The fundamental paradigm shift is the transition to twisted-pair as the connecting infrastructure. Because the electrical signals on twisted-pair must be regenerated by the hub, defective packets are not repeated to the nodes. Note that the collision (jam) signals are regenerated to provide indication to the originating node that its packet was not delivered. The hub does not know where the signals originated, and must propagate the jam signal. Simultaneous communication has been demonstrated for switched and duplexed packet delivery on Ethernet with collision detection disabled through specialized wiring hubs. Switched broadcast, multicast, and collision packets would be broadcast to all nodes, while packets destined for a single node would get sent only to that node in these specialized adaptations.

Figure 2.24 presents cascaded hubs for expanded network services not requiring extended node locations. Note that drop cables are limited to 50 m, whereas 10Base-T lobes are limited to 90 m, or 190 m with 100BaseVG. This stand-alone configuration also serves to demonstrate how easy it is to mix 10Base-T or 100Base-T twisted-pair with 10Base2 or 10Base5. At a fundamental level, Ethernet CSMA/CD is the same, regardless of the media. 100Base-T or 100BaseVG-AnyLAN match speed with 10-Mbps networks and also can be mixed with 10Base2 or 10Base5. The wiring hub—whether it is a 10Base5 fanout, or a 100Base-T, 100BaseVG-AnyLAN, or 10Base-T hub—saves on installation costs, simplifies maintenance, and adapts the transmissions between the faster and slower subnets. This design is easily transportable. It is useful, for example, to support trade show presentations, therefore providing an alternative to cables, taps, transceivers, and all the installation tools and time otherwise needed. The wiring hub is the fundamental connection, however, for supporting 100Base-T.

High-Density Service

Just as with Token-Ring or 10Base-T, the easiest solution to high-concentration Fast Ethernet nodes is the twisted-pair wiring hub or chassis that supports multiple hub cards. The simplest wiring hub will, for example, provide eight to 24 access points with twisted-pair for a tightly clustered office, easing the pressure on threading UTP wire long distances to meet the clustered need. This concept is consistent with

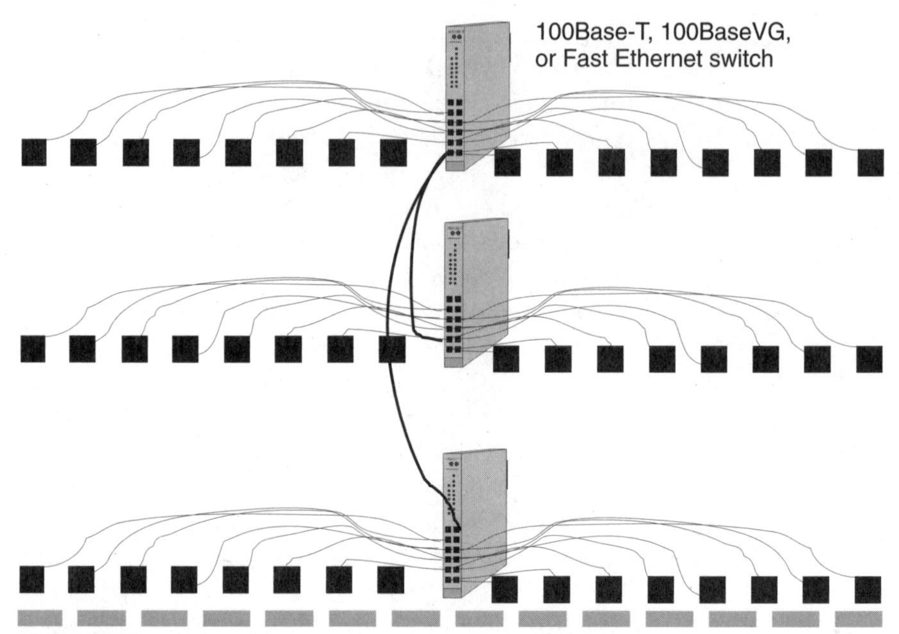

Figure 2.24.
Cascaded network configuration with wiring hubs.

installing workgroup hubs where the workgroups are and an enterprise hub at network central. In reality, the workgroup hubs are installed in the local telecommunication wiring closet, while the enterprise hub is located in the main network wiring closet.

Stackable hubs are hubs too, and are really no different from any wiring hub (other than vendor packaging differences). Larger chassis can support up to 18 modular cards, some of which can support up to 128 ports. All of these chassis also support Ethernet on fiber modules, FOIRL, FDDI, Token-Ring, redundant power supplies, and network management stations. Some additionally have modules for ISDN, ATM, X.25, frame relay, or T-1.

The use of twisted-pair hubs yields other economic and reliability benefits. The modular RJ-45 connectors are better than the Ethernet slide latch (used with 10Base5 and the T-connector used with 10Base2), and the integrated NIC units for twisted-pair have fewer parts. UTP wire is reliable, and so long as you qualify existing 10Base-T UTP for reapplication with 100BaseVG (which must be wired for four pairs) or 100Base-T (which must comply to Category 5 specifications), you can migrate from 10Base-T wire and connectors to 100-Mbps operation merely by changing network adapters, hubs, and software packet drivers.

If your network is based on a backbone coaxial technology, it is straightforward to add high-density service by attaching wiring hubs to the backbone. A standard transceiver cable connects each wiring hub. If a wiring hub replaced each workstation on a fully populated backbone, the maximum 100-node segment limitation, although still in force, could provide from 800 up to 4800 stations per segment, the higher number violating the limitation on Ethernet single-network addressing scheme. Furthermore, even 800 nodes would probably saturate the network unnecessarily. However, this example shows how high-density areas can be serviced easily as well as how to increase the effective network node access count with wiring hubs units. Economies of scale are usually achieved with twisted-pair wiring hubs.

The Bay Networks (formerly Synoptics) 3000 and newer 5000 chassis, for example, provides two-channel microsegmentation and ten half-width slots (including redundant power supplies) for cards of various types. When fully populated with modular cards, the network is limited to this double backbone architecture, as illustrated in Figure 2.25. It is important to recognize that the physical twisted-pair connection still functions and logically looks like the dual bus backbone network. Unless the A and B channels are explicitly interconnected, A is separate from the B network.

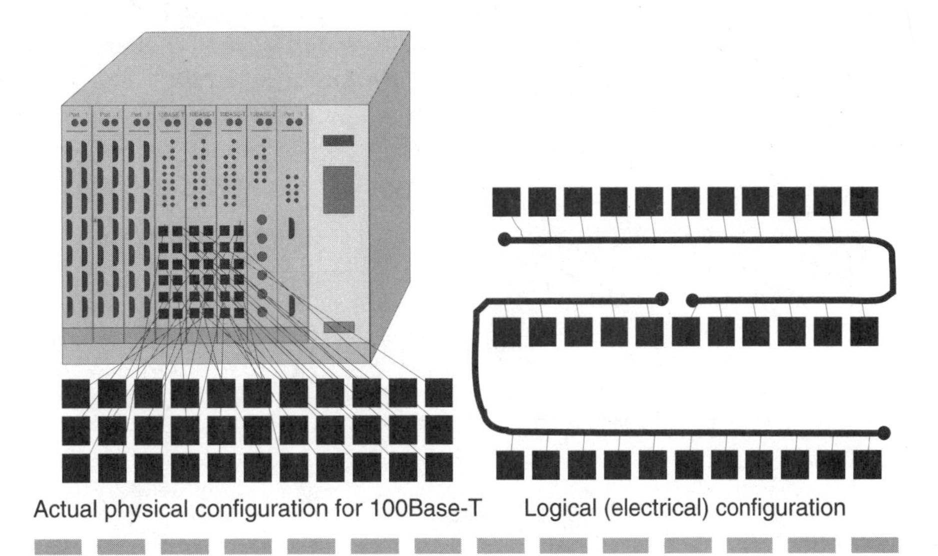

Figure 2.25.
Physical versus logical design with 10/100Base-T dual channel microsegmented backplane (dual bus backbone).

Network Bridges

The bridge connects different networks using different media together and promotes interconnectivity between multivendor networks. When you interconnect Ethernet with Fast Ethernet, you are most likely bridging media and transmission speeds. Bridges typically filter traffic at the MAC-level. Cut-through switches act like bridges, look like bridges, and so far, are the same speed if not slower than the fastest Ethernet bridges. On Ethernet, this reduces the probability for collisions, because a fraction of frames are filtered when addresses indicate that they do not need to be forwarded. Figure 2.26 illustrates these two principles. Only packets from nodes 4, 5, or 6 destined for nodes 1, 2, or 3 are repeated across this bridge. Likewise, only packets from nodes 1, 2, or 3 destined for nodes 4, 5, or 6 are repeated. Internetwork traffic thereby is reduced for an effective performance gain.

Bridges are also specialized hardware that translate transmissions from one type of network into that required by another. A bridge could also be a computer device that is compatible with two communication networks, interfacing with both. However, a bridge is totally transparent to the devices on the network. This means that protocols cannot be converted; however, NetWare running on Token-Ring can talk through a bridge to NetWare running on Ethernet if both are speaking 802.2 and IPS/SPX. Figure 2.27 shows a node that bridges two networks.

As larger organizations require enterprise networks and integrate microcomputer technology and microcomputer networks like Ethernet, the importance of the microcomputer grows. Users find the microcom-

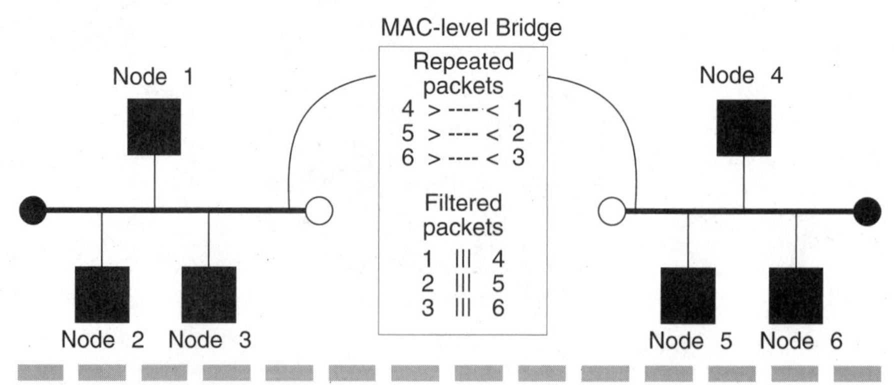

Figure 2.26.
A bridge filters traffic based on destination address.

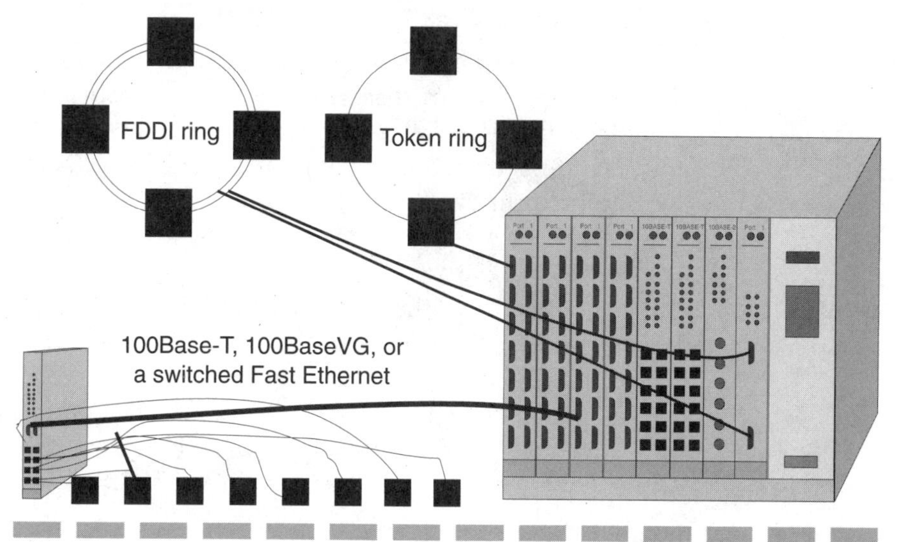

Figure 2.27.
An Ethernet is bridged to a Token-Ring network.

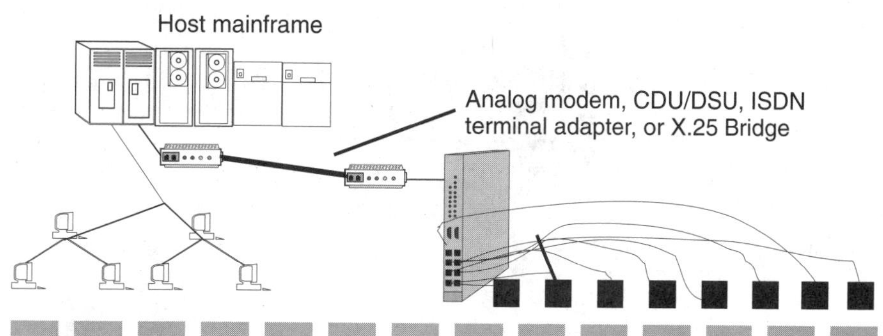

Figure 2.28.
A bridge from a mainframe to an Ethernet LAN.

puter less restrictive than mainframes, and more available. It provides software unavailable on mainframes. Networks based on file server technology providing client/server processing present a viable downsizing opportunity. Uploading and downloading of information between mainframe and microcomputer is a growing requirement, and bridges establish a common link, common transmission formats, and increasingly, common data formats. The mainframe-to-microcomputer bridge is explained by Figure 2.28.

Bridges provide access to layer 3 protocols with X.25, T-1, and networks like Tymnet and Internet. However, transmission protocols must decouple the packet transmission because transit time will exceed the expected Ethernet round-trip delay of 51.2 ms (at 10 Mbps). In some cases, such intermediate nodes will issue keep-alive messages to maintain the link; this is called *spoofing*. Devices such as frame relays or protocols including network block transfer protocol (NETBLT) encapsulate or interpret the frame for the linkage.

Note that a mainframe can be a node on Ethernet or Fast Ethernet when it talks that same protocol. Typically, however, mainframe protocols differ from the Ethernet protocol, and a router or gateway (something above the data link level such as a front-end processor) is required to translate or encapsulate these protocols for complex and mixed-protocol networks. Figure 2.29 shows an enterprise network.

Although bridges do support the expansion of networks into a network of networks, they can also limit that same expansion. Bridges must be fast, because they read every packet to validate and extract the MAC-layer destination address. Ethernet networks approaching packet saturation might overwhelm the capacity of a bridge to filter and forward

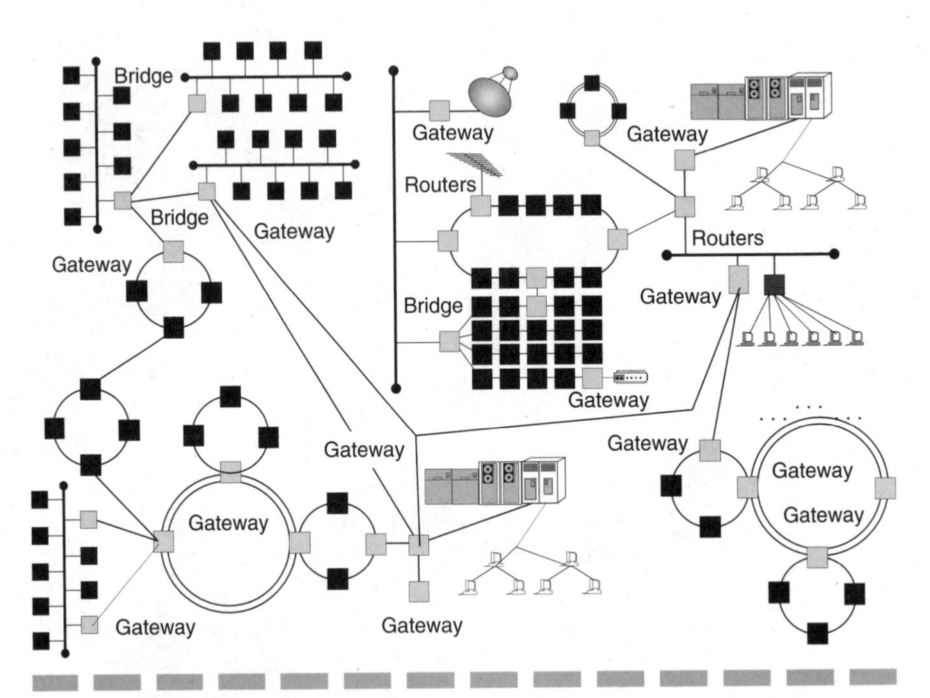

Figure 2.29.
An expanded enterprise network topology.

packets. If a bridge cannot accommodate the immediate network load and deal with it in real time, it will lose packets. Not only will performance suffer directly, the lost packets will have to be retransmitted for an added load. This traffic can create a bottleneck on the entire network and render it inoperable. Although bridges are transparent to network devices, the same is not always true for software applications. Bridges are not the solution for every network expansion. As an aside, every hub supporting multiple Fast Ethernet media and speeds is a bridge.

One bridging architecture very pertinent for Fast Ethernet migration strategies is the multiport density adapter. This device is typically installed in a server to connect into two or more subnets. This device is advantageous because it requires only a single motherboard slot regardless of the number of ports. There are also Fast Ethernet hubs that install inside the server. This is a useful configuration for small operations or workgroups, because it simplifies the hardware and wiring.

However, if you depend upon a multifunctional device and it is critical for operations, keep a spare around. It can be very difficult to replace a failed quad adapter with multiple cards and get the configuration working quickly. By using two or more physical and logical network addresses, some servers can increase their own throughput. However, this configuration is also useful for balancing loads over different subnets when limited network bandwidth is clearly the bottleneck. One of the more interesting quad-port adapters supports mixed 10/100-Mbps performance and also full-duplex transmission. This represents a good means to test migration or support mixed speeds. The photograph is shown in Figure 2.30.

Typical applications for this card include load balancing, subnet bridging, and experimentation with duplexed transmissions. Figure 2.31 illustrates several reasonable configurations. The first configuration shows a reasonable test bed for basic migration to 100 Mbps. The second possibility shows how to redistribute client nodes to increase available bandwidth. This is also a flexible method to migrate to 100Base-T without total replacement of all Ethernet hardware; start with just one side to upgrade. The third configuration shows construction of a high-speed backbone for server connectivity. Each server might be the central service facility for its own network, or it might share a central data repository. The high-speed and high-availability dedicated channel for the servers supports data synchronization, replication, and addresses one of the main performance issues with distributed PC-based client/server processing.

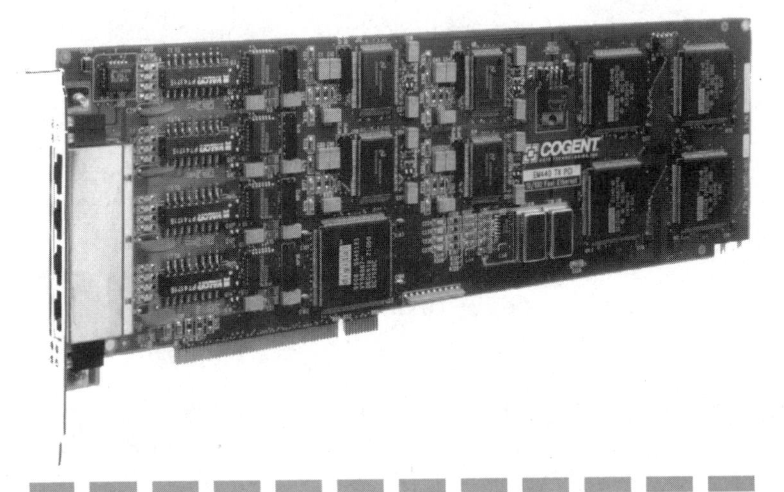

Figure 2.30.
Quad-port Ethernet PCI adapter supporting simultaneous 10 and 100 Mbps transmission rates. (*Courtesy of Cogent Data Technologies*)

Network Routers

A router is an intermediate node that is not protocol transparent. It does what a bridge does, except it can also convert some protocols for forwarding to other nodes. It must be addressed directly by a device on the network. Routers require routing tables that contain device addresses to identify network segments, the paths to these segments, and the relative efficiency of these paths. About twenty different routing protocols are in common use; each provides different service levels and interconnectivity benefits. A router does not use the tables to locate specific devices on the network; rather, it relies on other routers. A router uses this information (and the routing protocol) to determine the optimum route for each Ethernet packet, encapsulation, or spoofed connection.

The importance of routers when migrating from ARCNET, Ethernet, or Token-Ring to Fast Ethernet is twofold. Because most transitions are usually very gradual, preexisting networks will need to communicate with the new Fast Ethernet components. While a bridge can usually provide this service, routers are more robust in how they filter and route frames, optimize the path between networks of LANs, and translate different networking protocols. Specifically, routers provide a filtering component to improve network performance, an important concern when the migration to Fast Ethernet is driven by greater bandwidth and performance needs.

Routers offer the added capability to selectively filter traffic. Security for a network segment can be enforced by establishing a set of nodes that cannot traverse the router or be reached from the other side. Routers provide multiprotocol connectivity by translating or converting different protocols. Routers *tunnel* synchronous protocols (LANs are inherently asynchronous) such as SNA into a standard LAN format, and issue fake keep-alive messages (spoofs) to host front-end processors and end stations to maintain the impression of dedicated connections between host and end stations. Spoofing is increasingly important in remote LAN connectivity in order to minimize bandwidth requirements and telecommunication costs.

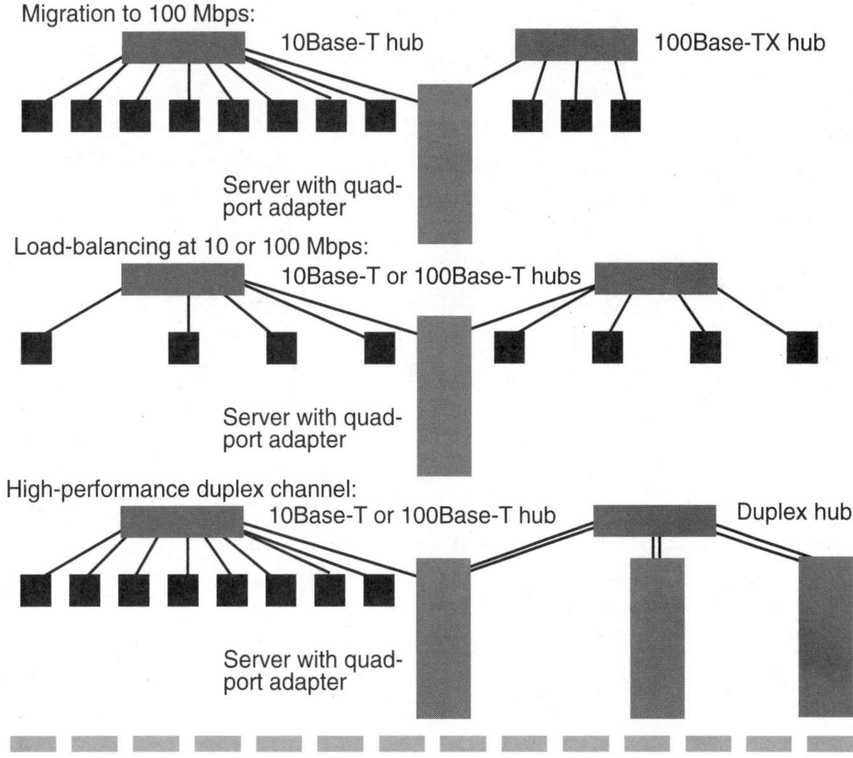

Figure 2.31.
Configurations for a quad-port Ethernet adapter.

A major differentiation between routing protocols is whether they are static or dynamic. Static routing requires explicit routing-table maintenance when devices are added, subtracted, or relocated on an internetwork, whereas dynamic routing automatically configures the tables. Dynamic routers usually require less maintenance, although they are inadvisable for large networks with many older network protocols because of the time required and traffic created when the tables are rebuilt and the routers *converge*. Nonetheless, routers form the basis for interconnecting the Internet, which is the largest network in the world, and it is the preferred interconnectivity device.

Routers are difficult to install, and require specialized knowledge and a high level of network knowledge for performance optimization. Routers (and enterprise hubs are often routers too) increase the complexity of network management dramatically, and make network failure troubleshooting more difficult. However, when bridges cannot maintain throughput because of overloads and buffer limitations, routers store and forward packets. Although routers are typically slower than bridges, they can handle a higher network load and are better than bridges at higher network loads. Routers are not the solution for every network expansion, for two reasons. First, the extra overhead associated with store-and-forward buffering adds to the processing latency. Second, they create traffic when they poll devices to maintain their automatic routing tables. Routing protocols and table maintenance represents a viable opportunity for optimizing routing performance, because many current routing protocols can unleash a broadcast storm of requests for active devices when they check the status of the extended network or converge on optimal routing paths.

Routers are one of the prime devices managed by SNMP management protocols, RMON agencies, and network management stations. They are good agents for assessing traffic, collisions, and timing errors. Some routers include priority network management station software and display SNMP statistics directly. Intelligent routing protocols can improve network performance through store-and-forward, routing cost or timing minimization, or prioritizing packet transmission by device address. Some of the newer routers include sophisticated filtering support that is useful for performance optimization, and also for constructing a secure firewall. Many routers will issue alerts to transmitting stations so they will not transmit (source quench) until the network traffic is lower. When router backlogs stifle the network traffic, the problems may be with protocol selection and application, network design, or network applications. Because most NOSs do not recognize this bottleneck for what it is, router overloads require personal intervention.

Spanning Tree Algorithm

Most Ethernet networks, in fact most networks, are configured so there is only a single path between any two devices. Even Token-Ring and FDDI networks provide only a single path between devices; the signal is unidirectional. (FDDI provides only a single path, which is multiplexed in the reverse direction only in case of primary ring failure.) Common network failures therefore can disable the entire network by breaking that single connection between nodes. Bridges and routers can partition the internetwork into still more functioning segments. However, this does not maintain network functionality for critical operations that might retain that functionality if multiple paths were available. Ethernet does not normally allow two segments to be connected with more than a single device, repeater, bridge, switch router, or gateway. Segments or subnets connected with more than one path typically introduce a routing loop. A routing loop (which is when you actually have multiple active paths) creates either a "black hole" where traffic simply gets lost, or a never-ending and typically growing traffic generator that eventually will saturate the network.

However, multiple bridges could be installed if they support the *Spanning Tree Algorithm* (*STA*). A spanning tree is any unique device-to-device path within the network, as Figure 2.32 illustrates.

Several spanning trees could exist between subnets. STA negotiates the optimum path and establishes one path as the forwarding path; all other paths are blocked. This guarantees a single active path between all node pair combinations so that there will be no active routing loops. Additionally, STA tracks for path failures and automatically establishes an alternative path in the event of failure. Note that because most spanning trees are enabled in routers, the router convergence time for establishing new routing tables and optimum routing paths is about five minutes per router on the network. As a result, the local networks may immediately be available, but 10 to 15 minutes could elapse before remote sites regain network access and service.

STA provides some redundancy for critical operations, but it is not the solution for immediately continuing fail-safe operations because of the delay for reestablishing routing tables; it can take up to 30 minutes to rebuild paths in router-based networks. Instead, use FDDI as the interconnectivity backbone for Fast Ethernet. It has comparable bandwidth, it is more robust than the Ethernet protocol at high load levels, and it reconfigures itself (that is, "ring wraps") within 35 ms when paths

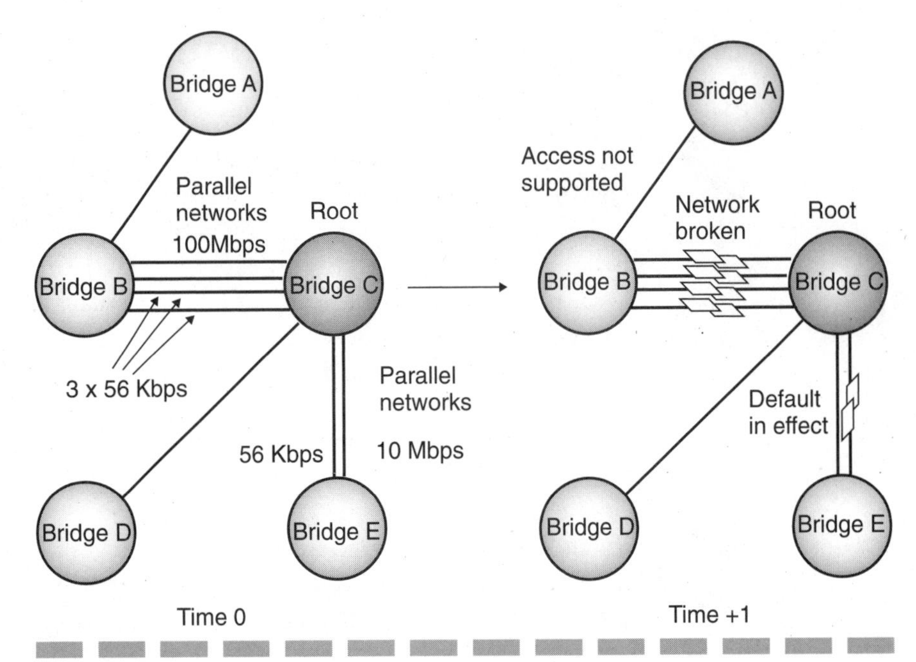

Figure 2.32.
Spanning trees provide redundancy without active loops.

are broken. That is faster than most Fast Ethernet switches. Many vendors provide FDDI hub adapters or FDDI connectivity as optional or basic Fast Ethernet enterprise hub functionality.

Network Gateways

Gateways interface between different network protocols and translate the transmission before retransmitting, as per the ISO definition. The gateway also has a software connotation, which is not to be confused with the ISO hardware definition. A gateway is often a connection between a host and a LAN, such as IRMA. More typically, a gateway is a software tool that provides interoperability functions at the OSI application and presentation layers. Such functionality includes data conversions, data presentations, and actual user file transfers.

A hardware gateway has two or more network access points to connect separate networks. The most usual form of a gateway is a workstation or computer processor with two Ethernet controllers, with each controller connected into a node on separate networks. This is how you might connect 10Base-T to 100Base-T, or 100Base-T to 100BaseVG, particularly because no vendor yet produces a mixed 100Base-T/100BaseVG hub. This is not a bridge per se, although this intermediate node itself performs a bridging function. This is a true gateway, because data and protocol interpretation occurs at all levels of the OSI model. The gateway unit supports only one Ethernet address on each subnet. The gateway is not a router per se, either, but it routes packets between subnets by resolving the destination addresses and rerouting the packets directed to it. Most computer workstations support multiple controllers, but require specialized software to function as a gateway (see Figure 2.33).

Network gateways and foresighted network planning can localize groups and functional units on different networks for significant performance gains. Although it is equally possible to subnet overloaded networks with either bridges or routers, typically the gateway handles excessive delays with ease. Data gets stored and forwarded. Such an approach is not effective with time-sensitive requests, mission-critical database applications, or on-demand imaging. However, it is a good solution to level traffic loads in bursty networks for synchronizing or replicating databases.

Subnetting isolates problems to each subnet and allows loads, demands, and storage requirements on file servers and print servers to be controlled by the people who use those services, averting the possibility of developing a politically tense situation. Subnetting isolates a group while lowering overall traffic. It is a solution to overloading problems induced by cascaded networks.

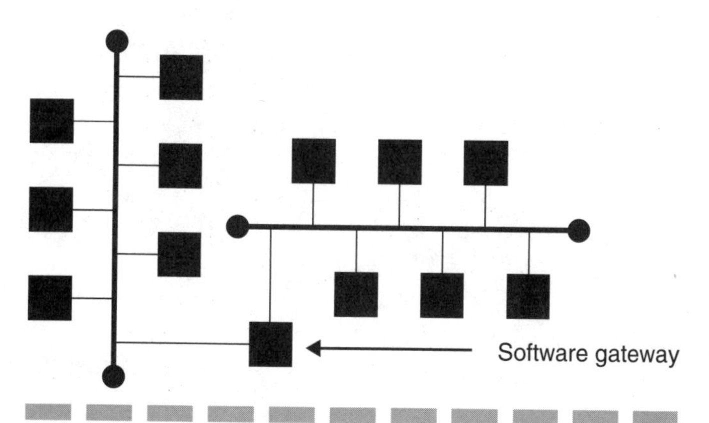

Figure 2.33.
A node serving as gateway to two subnets.

This is a manageable approach to controlling resources by group and division, as well as by individual subnets throughout the company. As an example, a backplane network would provide interconnection to all local subnetworks and supply network-wide access to expensive and low-usage specialized peripheral devices such as typesetters, image scanners, video entry cameras, pen plotters, coprocessors, and tape storage or disk storage devices. Subnets would concentrate each group onto a single Ethernet, so that excessive demand for services could be controlled, analyzed, and prevented from interfering with processing and traffic on other segments.

FDDI

Why even think about 100 Mbps with FDDI when Ethernet 100Base-T or 100BaseVG-AnyLAN natively supports 100 Mbps? The reason is that FDDI is the high-speed backbone technology that can nominally span distances to 50 km, and up to 200 km with full-lasing equipment. It is more robust than Fast Ethernet, and it is also more expensive to install and maintain. Many vendors provide interconnectivity for Fast Ethernet and FDDI. Nevertheless, this architecture is useful as a backbone for Fast Ethernet, as connectivity between singular points, or a fail-safe dual-homed connection for enterprise hubs. It is available today. Fiber Distributed Data Interface (FDDI) is proven. It is a dual ring, and therefore provides the fault tolerance within 35 ms that is required for ring wrap on primary ring failure. That is faster than any STA structure. (Recall that router convergence in a spanning tree can require five to fifteen minutes.) FDDI interoperates through wiring hubs, routers, or gateways. Although ATM, SONET, frame-relay, Fibre Channel, or the still-in-the-labs *dark fiber* promise faster backbone connectivity and WAN connectivity, FDDI is interoperable now.

It is important to recognize that the ANSI X3T9.5 standards committee drafting the FDDI definition in 1982 separated the physical layer into two subcomponents. This corresponds to the physical layer (layer 1) of the OSI model. These two components are the Physical Medium Dependent sublayer protocol (PMD) and the Physical sublayer protocol (PHY). For reference, the PHY component is logically positioned above the PMD component. Because it is only necessary that the PHY and PMD components know how to communicate between themselves, it has been possible to separate the hardware component (the PHY Physical sublayer) and redefine it to function on another medium. This allowed cabling developers to develop twisted-pair wiring to transport the FDDI signal over 100 m; this is called *CDDI* for *Copper Distributed Data Interface.* However, the electrical noise and RF pollution at 100 Mbps will pose a serious limitation in mixed use or residential neighborhoods. By the way, this is true for all 100-Mbps Fast Ethernet solutions on Category 5 UTP.

There are several fundamental applications for the faster 100-Mbps FDDI token ring technology to bridge or internetwork 10 Mbps or even 100-Mbps Ethernets. First, FDDI can serve as a simple backbone horizontal cross-connect or vertical conduit between Ethernet networks. It can also provide network and file server support and provide a more complex bridging function to interconnected Ethernet networks. When FDDI is used as a long-distance repeater for an Ethernet segment, its use must accurately match the Ethernet timing specifications, or excessive collisions will compromise the network. This is a fairly simple application for optical fiber.

FDDI is not the only optical fiber standard. Fast Ethernet with FOIRL is another option with fewer complications because the packet formats are the same; there is no packet encapsulation or translation. This is relevant when planning enterprise-wide networks, because the optical fiber should be compatible with all preexisting and planned Fast Ethernet segments. For more information about FDDI, refer to *FDDI Networking,* also by the author.

Fast Ethernet Logical Formats

Ethernet is defined as a *shared-bus* architecture. "Shared media" is the allocation of network bandwidth through the CSMA/CD or token-based protocols. This same protocol is true for all fast variants and is only bypassed by duplexed and switched architectures that create individual channels for each conversation. It is important to differentiate 100-Mbps Ethernet from 10-Mbps Ethernet, and duplexed or switched Ethernet from the standard shared Ethernet. You can boost Ethernet bandwidth performance with duplexed 10 or switched 10-Mbps Ethernet, or migrate to 100 Mbps, or enhance that to duplexed and switched 100-Mbps Ethernet. The logical structures are different. For example, shared standard Ethernet is limited to 10-Mbps bursts or some sharing of that bandwidth by all the nodes on the network. This logical architecture is true for 10Base2 or 10Base5, as shown in Figure 2.34. It is also true for shared 100Base-T and 100BaseVG.

A wiring hub is only a star-based physical redeployment of the same shared bus architecture. 10Base-T is still 10Base Ethernet, but it also includes signal regeneration that adds about 20 ms to each transmission delay. The shared media backbone is no longer a coaxial cable, but is a "backplane" that must conform within the same timing specifications of the coaxial-based LAN. Functionally, 10Base-T is the same as 10Base5, although media must be connected through a media converter. This is just as true for Fast Ethernet, as shown in Figure 2.35.

100Base-T and 100Base-T4 are also shared-media environments, but as Fast Ethernet, do almost everything 10 times faster than standard Ethernet. Because it is faster, it provides 10 times the base bandwidth within each interval. A 10Base-T network in constant collision may benefit from the faster clock times of Fast Ethernet, because collisions are resolved faster and there is that potential for more bandwidth. Even this Fast Ethernet will still saturate somewhere around 30 percent of capacity; however, this translates into 30 Mbps rather than 3 Mbps. As you can see in the next diagram, mixed Fast Ethernet hubs provide wider pipes to servers and other hubs, and increase the combined bandwidth avail-

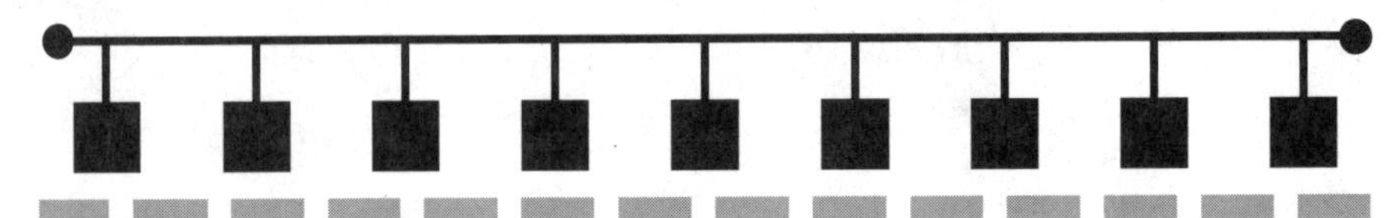

Figure 2.34.
Shared media means shared bandwidth.

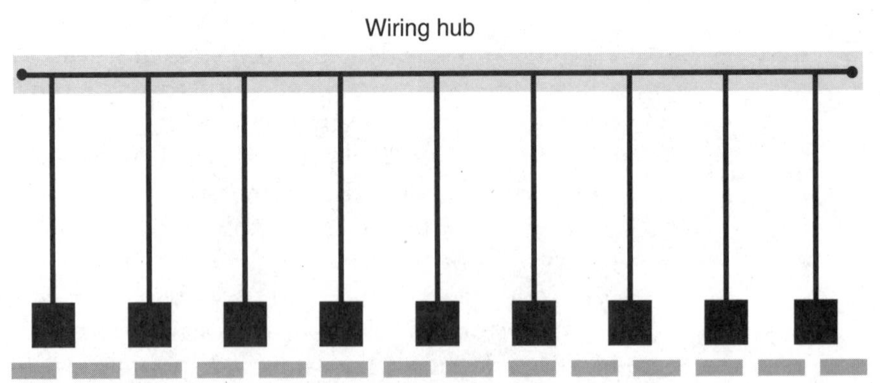

Figure 2.35.
A hub is just a shared media backbone in a box.

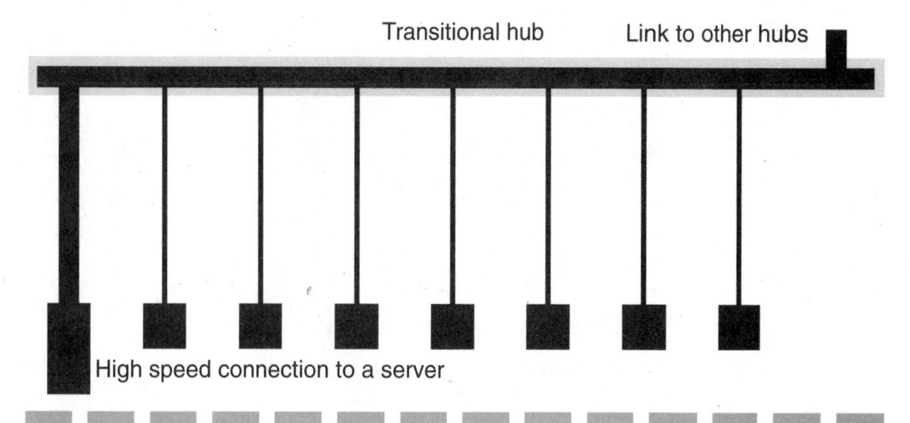

Figure 2.36.
Mixed Fast Ethernet hubs typically increase the available capacity for servers.

able to servers. This is useful for client/server networks overloaded at the network channel, as shown in Figure 2.36.

Although the hardware is currently expensive, some hubs support 100 Mbps to all devices. This is useful for bursty traffic (because of the clock rate increase) or for bandwidth-intensive processes. The hub is still a shared-media environment, and will saturate somewhere around 30-percent capacity, but the wider connections to every node reduce the chance for some types of client bottlenecks. Note that even Fast Ethernet is not immune to traffic bursts and collisions, so that you could encounter bandwidth saturation problems when you migrate from 10Base-T to 100Base-T. This is more likely if the network migration also corresponds to adding more users to the network, adding more applications, and increasing traffic loads. Notice that the logical diagram for 100Base-T looks just like that for 10Base-T except that the lines are wider, as shown in Figure 2.37.

The 100Base-T4 Fast Ethernet variant is still a shared media. Although it requires twice as many pairs, each of the four pairs only transmits 33 MHz of data. Each of four pairs is recombined into the 133-MHz (encoded) stream to yield the 100-MHz data rate. The network backbone is still the same shared media backplane, as shown in Figure 2.38, and does not provide any more throughout or performance than other forms of Fast Ethernet. In fact, to date, the 10Base-T4 cards and hubs perform about 25 percent slower than comparable 2-pair Fast Ethernet hardware.

The Hewlett-Packard 100BaseVG initiative is actually a different logical structure from other implementations. Although it is physically

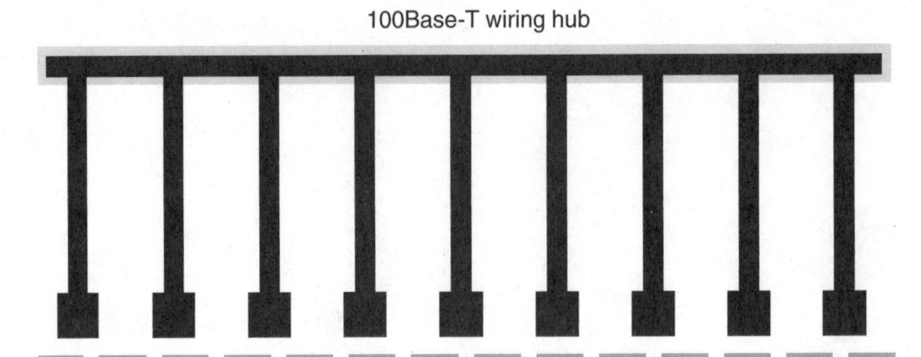

Figure 2.37.
Fast Ethernet still retains the shared-media heritage and its limitations.

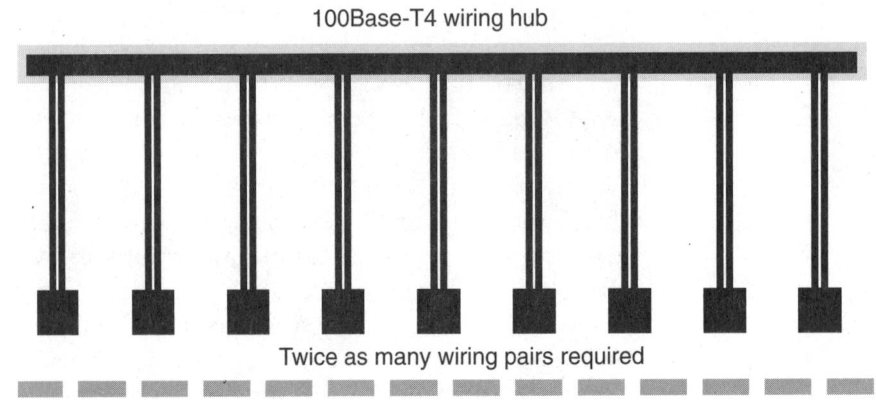

Figure 2.38.
100Base-T4 is still a shared media limited to 100 Mbps.

wired with NICs to a single hub, the logical structure is a ring. While the physical protocol retains the same Ethernet packet format, media access is controlled by a central round-robin mechanism. It is analogous to Token-Ring, but there is no token and the hub controls the access. This is illustrated in Figure 2.39.

There are three fundamental differences between 100BaseVG and all the other Fast Ethernet variants. First, the 100BaseVG lobe wire length can be as long as 200 m. Second, 100BaseVG (with the VGAny-LAN protocol) supports integration with Token-Ring and packet sizes to 18,543 bytes. In contrast, Ethernet packet sizes are limited to 1518 bytes. Third, this round-robin protocol trades the inefficiencies of a collision-detection for the determinacy of a token-like protocol. The tradeoff is that the minimum 100Base-T signal transmission time is about 1 ms, while the 100BaseVG hub circulation time ranges from a minimum of 20 ms to an average of 50 ms. That is only about as fast as 16-Mbps Token-Ring and standard Ethernet. However, while standard Ethernet can saturate and yield less or even no data throughput, demand priority is robust and guarantees delivery even under the most burdened traffic levels.

There are two methods to bypass the limitations inherent in the Ethernet protocol's collision-detection without changing the protocol itself. These methods include duplexed transmissions and switched architectures, as previously mentioned. Duplexed transmissions provide two one-way connections between the shared media. In other words, there is a directional backbone for traffic one way and another backbone for traffic the other way. On a network with 20 Mbps, the realistically sustainable traffic level is less than 16 Mbps, still about a five-fold increase over standard Ethernet performance. This architecture, as shown in Figure 2.40, is effective in busy client/server networks because while servers are usually bottlenecked by a series of disk I/O and network I/O traffic, they can usually own the outbound transmit channel and constantly keep it full.

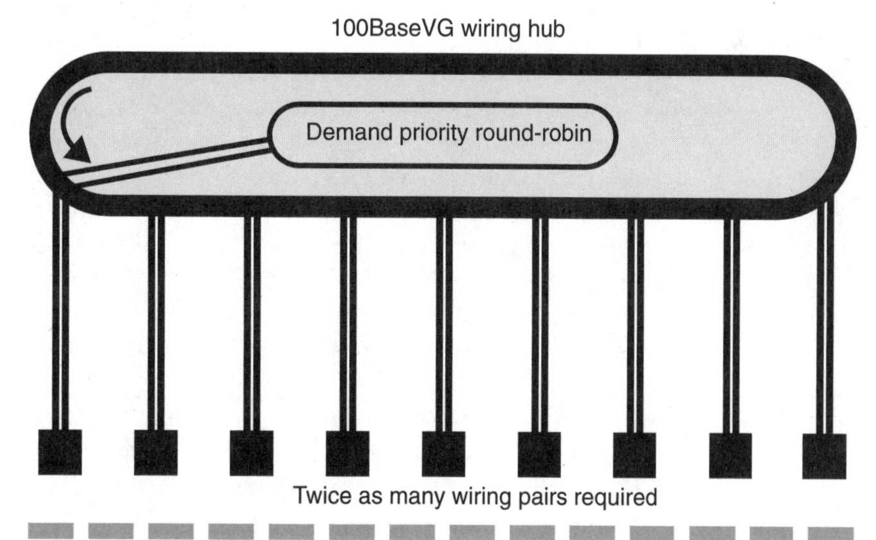

Figure 2.39.
100BaseVG used a demand priority protocol.

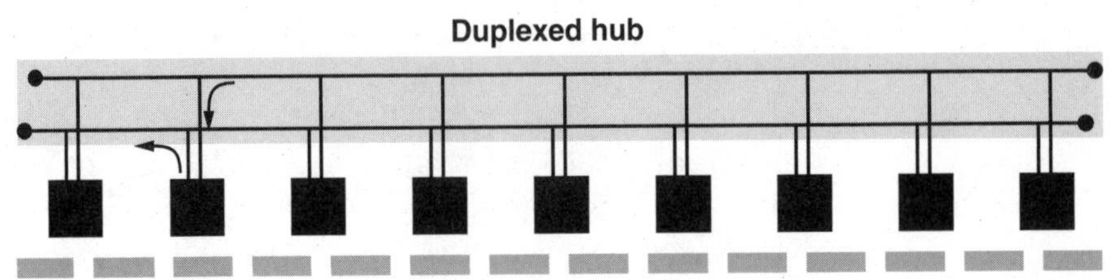

Figure 2.40.
Duplexed connections are effective with structured traffic loads.

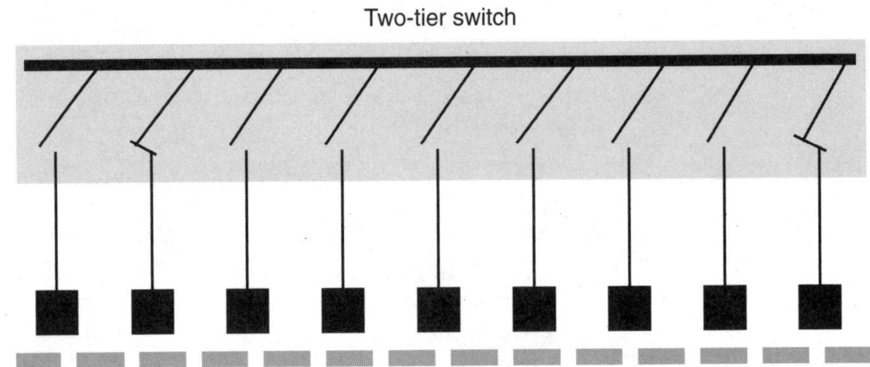

Figure 2.41.
Switched connections are the ultimate in bandwidth microsegmentation.

A switched network is the ultimate in network microsegmentation, because every transmission is sent on its own switched private network. A backplane handles many gigabits of simultaneous conversations. Nominal bandwidth on a 10Base-T network with 10 switches is 50 Mbps. It is not the 100 Mbps you might expect, because devices on the switched connections tend to talk to each other. Realistically, each switched connection requires from 10 to 600 ms to establish, and nodes trying to talk to the same server at the same time will create a timing bottleneck. Switched performance works best with a peer-to-peer network, as shown in Figure 2.41.

Switched networking will ultimately provide the greatest bandwidth. Selection of hardware and the layers of intermediate nodes in the network architecture play a critical role in the success of the switched variant of Fast Ethernet. Older switches are not much different from routers, and provide the same slow performance characteristics. Newer switches, particularly those switches with ASIC hardware, perform much better.

Fast Ethernet Designs

In addition, the mechanical standards also bend the standards from the original single one-way conversation (*half-duplex*) to simultaneous two-way (*full-duplex*) conversations. Figure 2.42 illustrates this important difference.

Although standard Ethernet provides 10 Mbps in capacity, traffic loads, delays caused by Ethernet busy signals (recognized signal on the network), and normal collisions easily reduce this to an effective throughput of 1.25 Mbps. Different combinations of network device counts, packet sizes, requests for transmissions, and actual packet transmission frequencies result in throughputs as low as 125,000 bps or as high as 9.4 Mbps, which is the theoretical capacity when all overheads are included. Ten workstations or thirty PCs with a single server can easily saturate that capacity. Internetworks with hundreds, thousands, or the handful of networks with 100,000 and more devices simply could not exist within this *bandwidth limitation.* The original Ethernet design included bridges and routers to support segmentation of a single network

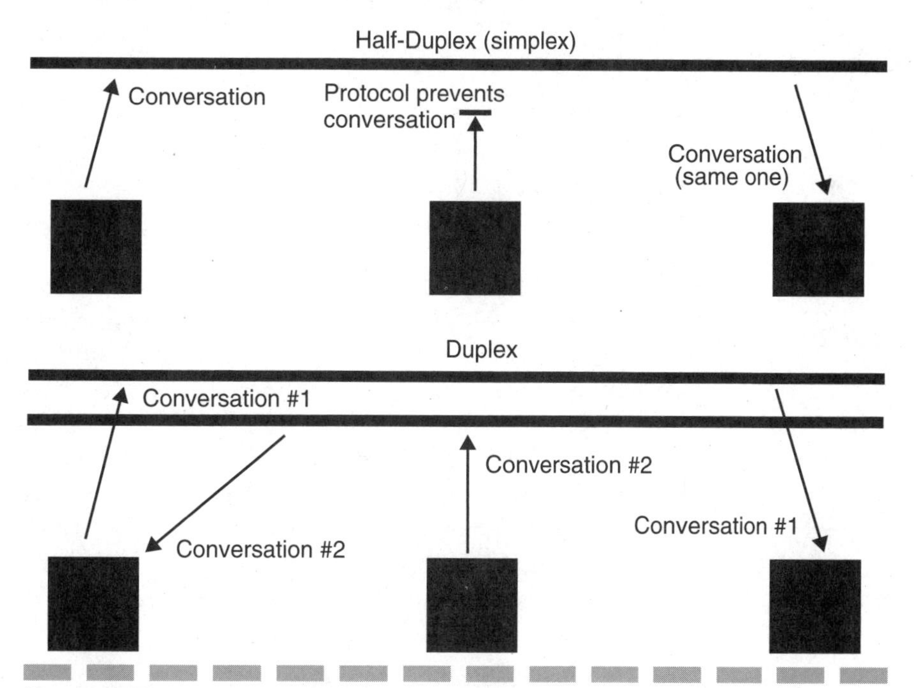

Figure 2.42.
A comparison of half-duplex and full-duplex transmissions.

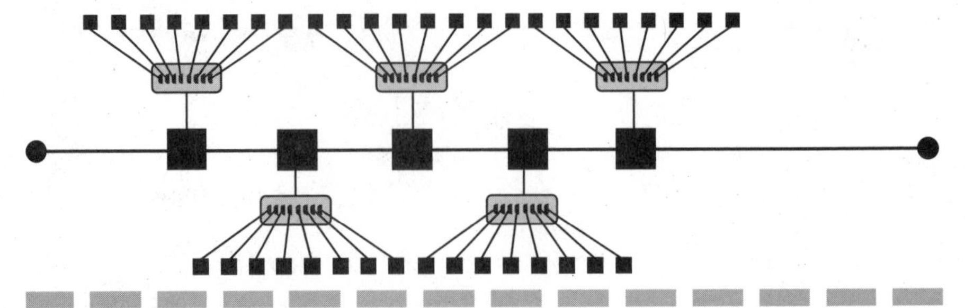

Figure 2.43.
Segmentation of a network into subnetworks.

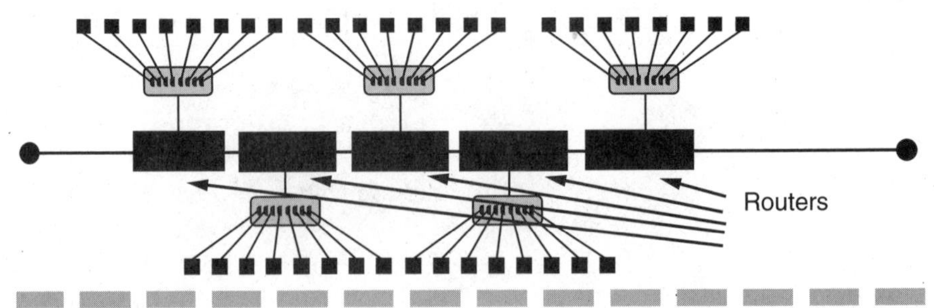

Figure 2.44.
Ultimate router-based microsegmentation.

into logical subnetworks that are still connected, in order to break apart a large network into something supportable and manageable. Figure 2.43 illustrates this architecture.

Because the network is segmented into smaller and smaller subnetworks, this technology is called network microsegmentation. The ultimate *microsegmentation* is the impractical subdivision of a routed backbone with a single node on each subnet or router for each hub, which is shown in Figure 2.44.

This architecture often exists in organizations—it often grows over time—and can represent a significant performance bottleneck because all the traffic among routers must travel the backbone. It is possible to create router mesh to optimize traffic patterns. Another fundamental problem of segmentation is that transmission time between devices is increased by each intermediate layer. Sometimes even the 1.25 Mbps shared among the devices on each subnet is still insufficient to accommodate the traffic loads with a handful of workstations. As a result, a switched segmentation was designed to provide full protocol bandwidth for pairs of nodes on-the-fly. Router-based hubs replaced simple wiring hubs to make real the impractical one-for-one router architecture. As stated previously, the monolithic backbone can provide bandwidth greater than Fast Ethernet at widths up to 10 GB. These routed hubs are real products. Such products are generally called workgroup hubs, enterprise solutions, or some other glib product names. The actual architecture is illustrated in Figure 2.45. Cisco logically models the actual backbone (the "black box" in the illustration) as a series of rotaries; other vendors show meshes or fat pipes. Performance is a function of the routing hardware, the routing protocol-whether store-and-forward or pass-through, and the sustainable load that the hub and the routing backplane can sustain.

Although these routers included built-in communication channels that were in themselves backbone networks with bandwidths in excess of 400 Mbps, these *backplanes* were limited by the routing performances of the hub architecture. Routing times range from 300 to 600 microseconds (ms), at least a magnitude greater than the basic Ethernet transmission times. Routing times are part of the transmission delays called *latency*. Latency is a serious performance bottleneck discussed in greater detail at the end of this chapter and referenced throughout this book.

When it became clear that networks represented a growth industry, that microsegmentation was useful for resolving bandwidth limitations, and that the discrepancy between 2-ms packet transmission times and 600-ms routing times provided substantial business opportunities, vendors moved the routing software into hardware to create network switches. A *switch* is just a router built in hardware rather than on a standard PC or SPARC platform. The hardware is a specialized chip (RISC or ASIC) integrated into a bus supporting massive communications, and it yields the type of microsegmentation as shown in Figure 2.46.

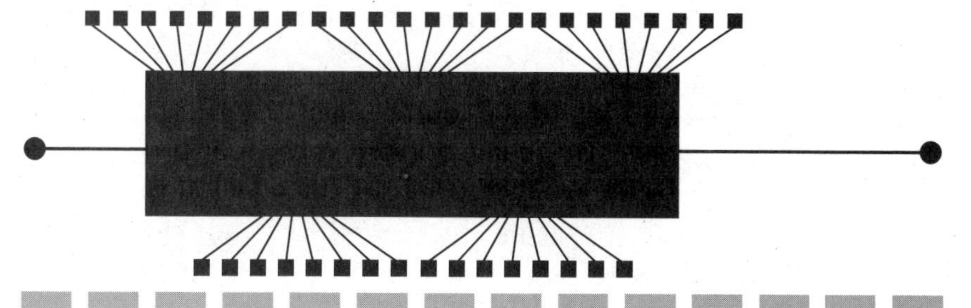

Figure 2.45.
Router-based hub segmentation.

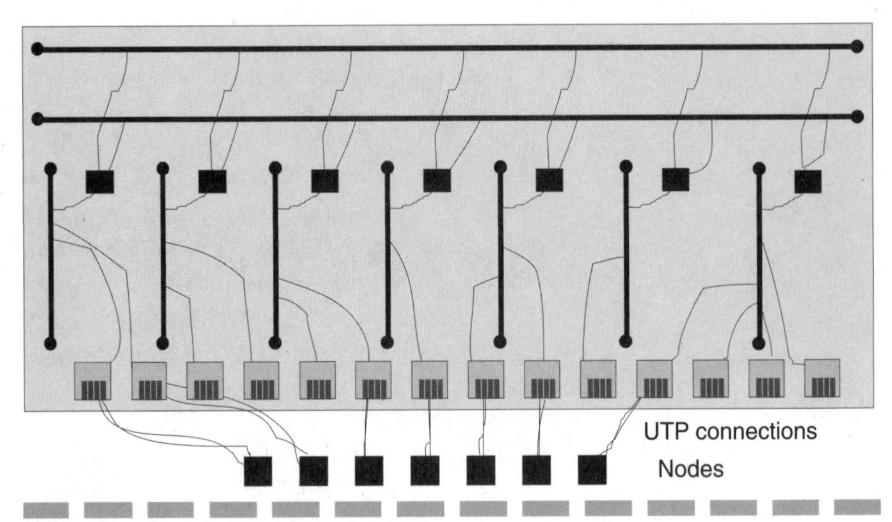

Figure 2.46.
A switched architecture is the ultimate in microsegmentation.

Virtual LANs, or *VLANs,* are a wiring and design construct only. Although they are switches supporting Ethernet and Fast Ethernet, they do not increase bandwidth, provide microsegmentation capabilities, or minimize latencies. VLANs do not increase transmission speed. A virtual LAN is a software-defined group of devices that communicate as if they are on the same LAN. It is only a wiring configuration used to create different combinations of subnetworks. A VLAN is entirely a physical implementation of the logical design. Think of the prior discussion of the logical network designs, and imagine converting them into a physical layout; that is what the virtual network switch does. In a way, it is too bad that switching and virtual switching have similar names. It's confusing.

Virtual switching is easily accomplished by rearranging the lobe connections in a central wiring closet. This rearrangement, and that is all it really is, becomes *virtualized* when the patch panel is automated with a switching matrix and remote software. Think of a VLAN as the public phone system extended to data networking. When the population of devices on each subnetwork is load-balanced, you can achieve some performance benefits. For the most part, this technology is outmoded by switching. By the way, software recognition of where devices are in relation to servers, routers, and bridges is a more complex problem, one not totally resolved by the current network operating systems. Figure 2.47 illustrates this logical/physical architecture.

Media

Ethernet was originally transmitted over a *coaxial* cable, which is a single-strand cable with a tinned copper core, surrounded by a foam material that insulates it from the tinned and braided copper shield. Coaxial cable is often used because it provides an electrically balanced signal that overcomes the problems of crosstalk, signal self-destruction through attenuation, and high impedance. It is also expensive and unwieldy. AT&T and Synoptics figured out how to extend Ethernet to standard telephone wiring. StarLAN and LattisNet became the basis for all the subsequent Ethernet variants and the realization that 10 Mbps could be separated from the media and enhanced.

Unshielded twisted-pair (UTP), also known as *datagrade* wire when used for networking, is the prevalent medium for Ethernet and Fast Ethernet. Twisted-pair wiring forms the underlying physical connectivity for 10Base-T and 100Base-T Ethernet variants. I mention the use of datagrade (Category 4 and higher) wire rather than generic voice-grade (Category 2 and lower) telephone cable because there are many grades of unshielded and shielded twisted-pair, and cable that is rated for data applications is far more reliable and flexible in terms of future applications. This is illustrated in Figure 2.48.

The most astute choice for twisted-pair wire in new installations is generally defined as EIA/TIA 568 Category 5 unshielded twisted-pair (UTP). This wire is rated for data transmission at speeds up to at least 100 Mbps with proper installation. Some companies, Belden for example,

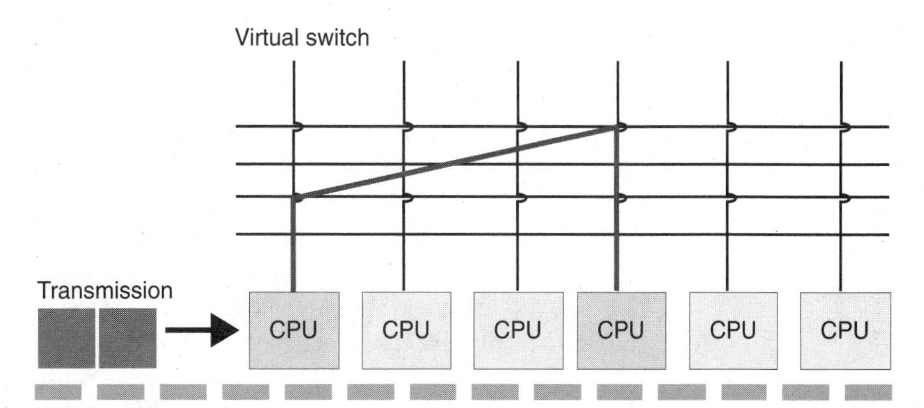

Figure 2.47.
Virtual LAN architectures are the physical automation of the logical microsegmentation and represent few performance enhancements.

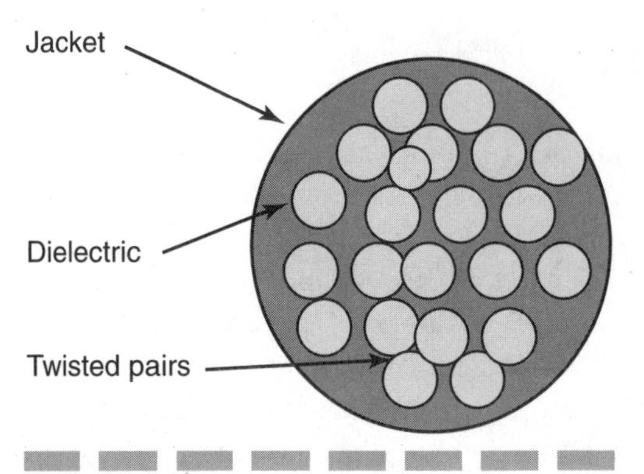

Figure 2.48.
Unshielded twisted-pair cable contains multiple bundles of two wires each that are a set, and that set is twisted together.

advertise performance characteristics up to 350 MHz for their Category 5 wire, perhaps suitable for 175 Mbps or even 225 Mbps with an MLT-3 or quartet signaling encoding scheme. Belden and EIA/TIA are also discussing a Category 6 material. Although present-day Category 5 performance requirements might not exceed 155 Mbps (with at least one Fast Ethernet product), use of this material provides future flexibility for use with asynchronous transfer mode (ATM) on two pairs of copper.

Coaxial cable consists of a conducting outer metal tube enclosing and insulated from a central metal conducting core. The signal is actually carried only by the inner core, and is shielded by the outer tube, which acts much like a radar wave guide. Although there is a voltage differential between the conductor and shield, do not be fooled that coaxial cable is wire and that one-pair wire can substitute for it; coaxial cable is a specialty product. It is called *coaxial* because the core and shield conductors share a common axis. The cross-section for coaxial cable is shown in Figure 2.49. Coaxial cable is not particularly important for Fast Ethernet as a medium, because the specifications and vendor products do not support Ethernet speeds greater than 10 Mbps. However, 10Base2 or 10Base5 coaxial often serves as a backbone to connect Fast Ethernet workgroups with light internetwork traffic needs such as E-mail and messaging. If the traffic load can be contained with the workgroups, there can be minimal traffic loads between Fast Ethernet segments. In fact, some low-end 10Base-T/100Base-T hubs also include a BNC connector.

If you need to use FOIRL or FDDI for internetwork connectivity, that bandwidth requires optical fiber. Optical fiber is a glass or plastic tube that carries signals on laser light at wavelengths of 780, 850, 1330, 1550, or 1610 mm. Transmission is digital transmission from a light-emitting diode (LED) or laser over the core. Fiber is unidirectional, so most networks typically require twin fibers. A single 100Base-FX or FDDI fiber is shown in Figure 2.50. Optical fiber is smaller and lighter than UTP or coax, because it typically contains no metal. Armored fiber (with twisted cables or wrapped galvanized sleeves) for underground installations is as large as a garden hose, and is very stiff because of the metal reinforcements. The armor requires special grounding to prevent electrical or lightning conductivity.

Any concerns about the difficulty of using fiber, the costs of fiber, and fiber termination can be easily resolved by buying ready-made jumpers and cables. Under those circumstances it is just like using Category 5 patch cables. Otherwise, you will need cutting, polishing, crimping, and connector-gluing tools. If the fiber is exposed or you rely on many segments of optical fiber, you might need test equipment that parallels what you have for copper cabling. The real disadvantage with fiber is not the cost of the fiber, but rather the cost of the diagnostic tools. This in part can be offset by buying used tools for fiber. There are a number of vendors recycling early AT&T test equipment.

Optical fiber suffers transmission degradation over time through exposure. Expansion and contraction caused by extremes of heat and cold fracture the cable. Heat, cold, and particle radiation discolor the optical qualities and gradually decrease the bandwidth. Plastic cable is more durable than glass-core fibers, but does not yet provide the same

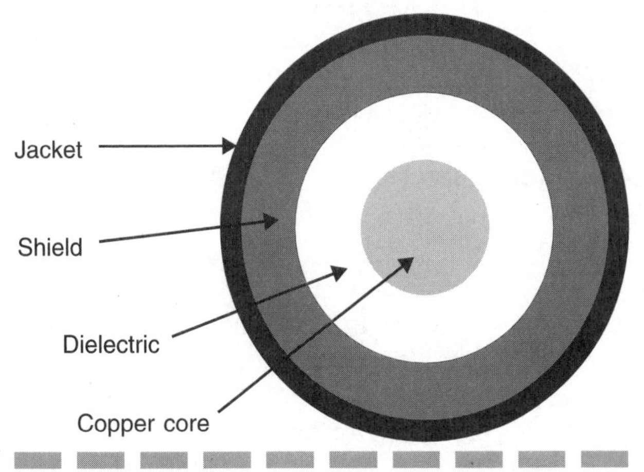

Figure 2.49.
Coaxial cable is tube within a tube.

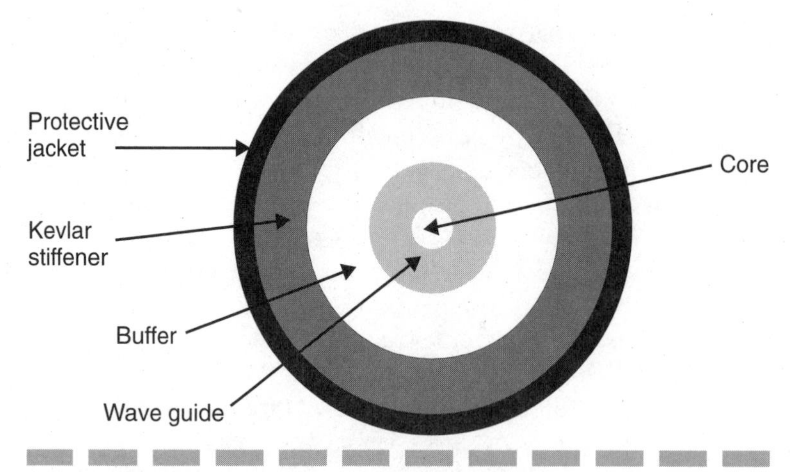

Figure 2.50.
Fiber-optic cable is usually nonmetallic and unidirectional.

clarity as glass. Managers at several sites with very large installations (generally with nuclear power generation, R&D labs, or nuclear medicine services) have discovered this firsthand. Nevertheless, the logical network designs later in this chapter show that fiber supporting a 100Base-FX linkages to a central enterprise hub or a promiscuous FDDI backbone is an ideal method to interconnect larger sites.

New installations are usually established as twisted-pair sites due to the obvious wiring simplicity, widespread support, and low entry price. The main differences between Fast Ethernet variations are the costs, the media, and the maximum lobe length. If you select 100BaseVG, interconnectivity and interoperability with existing networks is a little harder to accomplish, but otherwise very possible. See Figure 2.51 for Fast Ethernet media variations.

All variants support the IEEE 802.3 versions. Note that there are differences, and all these variants are unlikely to coexist successfully on the subnet or backbone without a physical media and data link bridge between them. The ANSI and ECMA standards are acceptances of the IEEE standards, often without any changes from the original standard wording. In the same way, you might note that many standards might bear the designations from multiple organizations in the form of just EIA/TIA (or the ridiculous EIA/TIA/ICFA/ANSI/NIST/ISO, which confounds hyphenation dictionaries), thus representing a concerted effort to rationalize the different and competing efforts to set international standards.

To sum up the media limitations for Fast Ethernet, consider the following. Fast Ethernet runs readily over shielded twisted-pair, unshielded twisted-pair, and sometimes voice-grade or legacy phone system wire. In all cases, reliability is enhanced with an infrastructure built from data-grade wire and backed up with graded connectors, punch-down (patch) panels, and quality installation.

For practical migrations, you typically can recycle both 4- and 16-Mbps Token-Ring installations to 100Base-T. Some newer 16-Mbps Token-Ring installations were wired with UTP that corresponds to Category 4. Unless four pairs were wired from the telecommunications closet to the user, you will have to install new cable for 100Base-T4.

All twisted-pair hub to workstation connections are *nominally* limited to 100 m, or 200 m with 100BaseVG. In reality, the lobe cable is limited to about 90 m with a workstation jumper of 3 m, and a wiring closet patch cords of 6 m; 100BaseVG supports 190 m of lobe cabling. Typically, the lengths of all cross-connects (as the jumpers and patch cables are generically called) should be 10 m or less. Some hub vendors regenerate and reclock signals to the node for effective service to lengths that exceed 130 m; however, proceed with caution.

Design Limitations

The fundamental limitations of Ethernet are driven by signal-timing issues, primarily collision recognition times. This means that 500 m is the limit for a coaxial network backbone, 50 m for transceiver lobe cables from that backbone to each network device, and 100 m for UTP lobe wiring. Fiber extends lobe lengths to 427 m and can connect backbones up to 2 km distant with point-to-point connections. The Ethernet bus does not need to be linear, and can wrap in many shapes for

Specifications	100Base-T	100BaseVG	100Base-FX
Media	Twisted-pair	Twisted-pair	Optical fiber
Topology	Star wiring	Bus/star wiring	Bus/star wiring
Nodes/subnet	100	200	100
Network length	100 × 2	200 × 2	427 × 2
Lobe length	100	200	428
Bandwidth	10/100 Mbps	10/100 Mbps	10/100 Mbps

Figure 2.51.
Fast Ethernet media variants.

greater coverage. Contrast these logical configurations with a building floor plan shown for scale. The floor plan shows a building 94.5 × 39 m (295 × 122 ft) for contrast; it is the same floor plan shown throughout this book as the characteristic floor in an office building. This is illustrated in Figure 2.52.

Each Standard Ethernet backbone can be attached by fiber links of length to 2 km each. These links must be point-to-point; you cannot interrupt them for user nodes or other equipment (without inserting other repeater links). The floor plans are again included to show the scale of standard Ethernet coverage in Figure 2.53.

The limits for Ethernet, and thus Fast Ethernet also, are two connections between any three subnets that communicate together. Each subnet is fully addressed LAN entity. Although it is possible to create a much longer nonlinear chain, this architecture prevents subnets from communication with other subnets. However, you can create a central subnet that is the hub for all (almost any number) of the subnetworks. Note that this limitation creates a maximum of five levels for hubs, concentrators, or other cascaded architectures. The three subnets connected by two fiber links are illustrated in Figure 2.54. The floor plans are again included to show the scale of standard coverage.

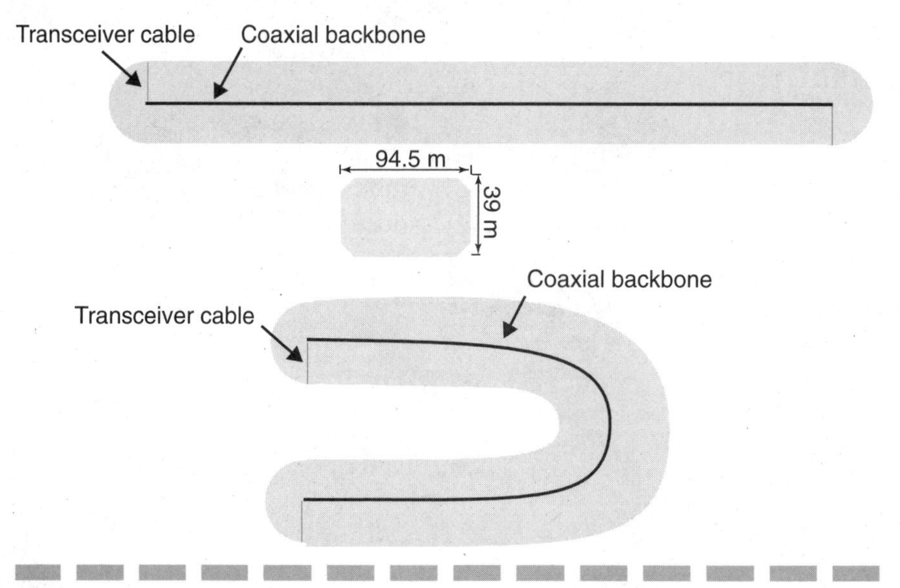

Figure 2.52.
The 10Base5 coverage range.

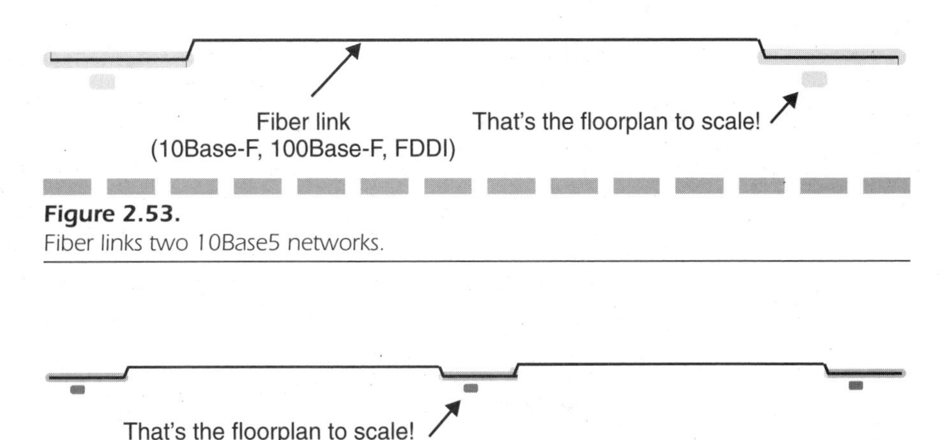

Figure 2.53.
Fiber links two 10Base5 networks.

Figure 2.54.
Fiber links three 10Base5 networks, the maximum architecture.

Although 10Base5 might seem irrelevant to Fast Ethernet, consider these basic design limitations when migrating slow networks to faster ones. Because coverage is different, some sites might require routers and fiber links to do what you are presently doing with standard Ethernet. The needed migration changes could easily create performance bottlenecks and implementation problems. Contrast 10Base5 to the new UTP Fast Ethernet wiring ranges in Figure 2.55.

However, if you'll recall, the two-fiber-link limitation imposes a cascade level limit of five deep. This same limitation persists with Fast Ethernet. However, you can create an enormous coverage area with 100Base-T or 100BaseT4, and double that for 100BaseVG, as shown in Figure 2.56. The floor plan is again included to show the scale for Fast Ethernet coverage.

Realize that these logical designs show extended area coverage. Although you might actually create such designs for real applications, fully populating all the hubs could create a network bottleneck even at 100 Mbps because the aggregate from all the nodes (as few as 64 and as many as 628) might saturate the hub-based backbone. When you must reach into far-off corners of large buildings, install a fiber linkage. This works as well for 100Base-T as it does for 100BaseVG, although the logical design in Figure 2.57 shows the maximum number of linear linkages. The floor plan is again included to show the scale for Fast Ethernet coverage.

Fiber to the desktop promises greater bandwidths than just 100 Mbps. Although this is the functional limit for both Fast Ethernet and FDDI, new technologies such as Fibre Channel, dark fiber, and Ultra Fast

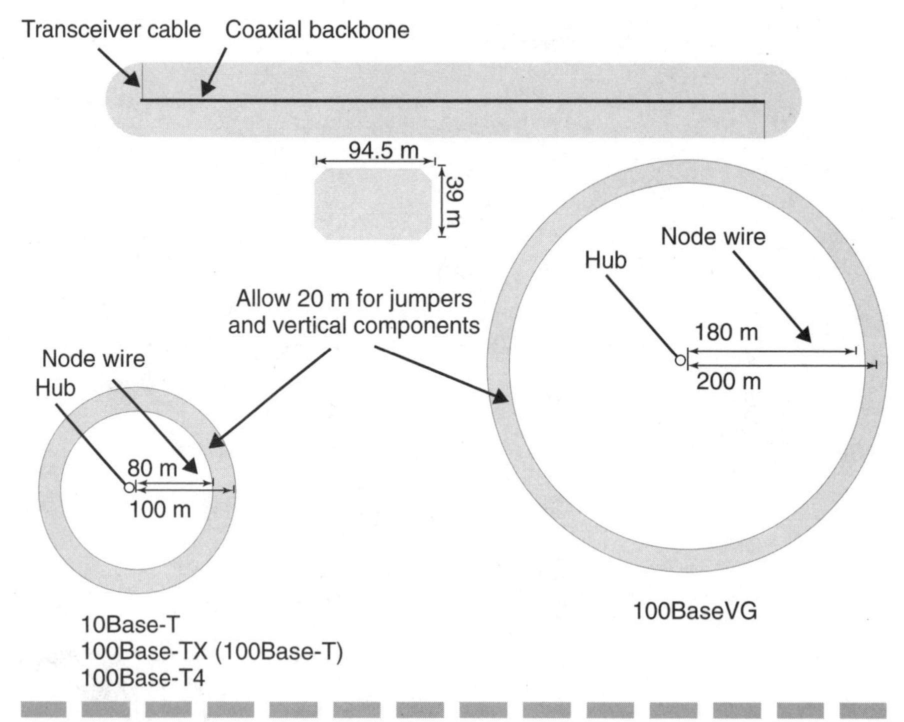

Figure 2.55.
Fast Ethernet wiring coverages.

Ethernet will exceed these bandwidths. Although there is no functional difference between 100Base-T and 100Base-FX in terms of speed and performance, fiber does provide several benefits worth considering. FDDI is usually no better and often more complicated than 100Base-FX. FDDI is more effective as a backbone to interconnect busy LANs. In that case, the deterministic performance of FDDI overcomes the Ethernet saturation problems under extreme loads. The first benefit for any fiber connection, as shown in Figure 2.58, is the increased lobe lengths between a hub and the network devices, which are even greater than the length of the standard Ethernet coaxial backbone.

Second, fiber is difficult to tap without detection, immune to electrical noise, and more reliable in most environments than UTP. In fact, fiber costs about the same as shielded twisted-pair (STP) or shielded foil twisted-pair (SFTP) for materials and really is the same cost installed as Category 5 UTP given the material shortage and currently inflated prices for UTP. NICs and fiber adapters are more expensive, however. If you are looking to the future, fiber infrastructure should be easy to migrate to faster fiber-based protocols such as dark fiber, Ultra Fast Ethernet, SCSI-3 (SC-FC), or Fibre Channel.

One of the common mistakes I see in 10Base-T, 100Base-T, and other copper installations is that designers measure wiring paths in terms of linear distances. They typically forget the risers, the heights, and the actual paths wire lobes take between a hub and a node. They also forget the jumpers and patch cords in the telecommunication closet. Notice that all the prior logical diagrams included a two-tiered ring to show specifications with practical installation limitations.

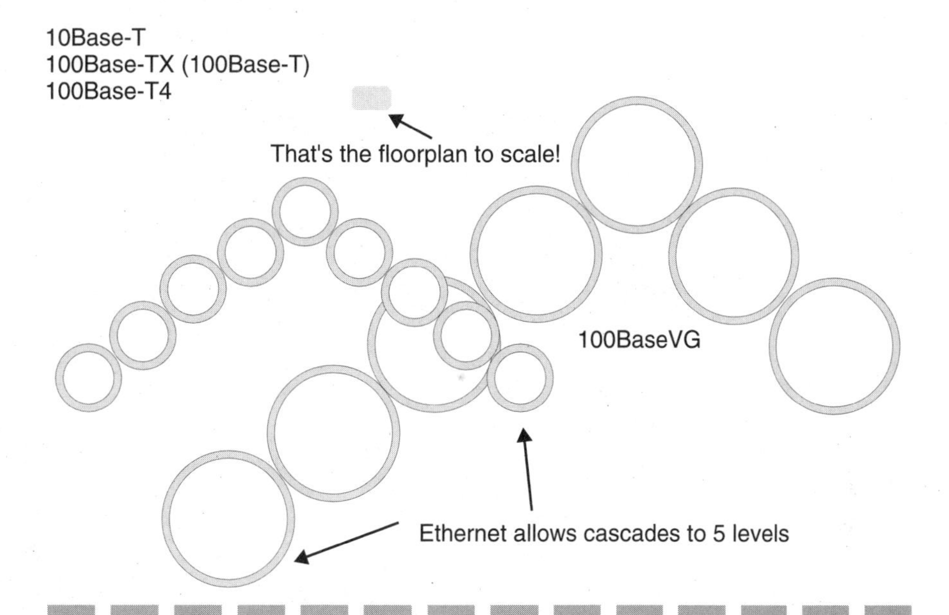

Figure 2.56.
Fast Ethernet wiring coverages with hub cascades.

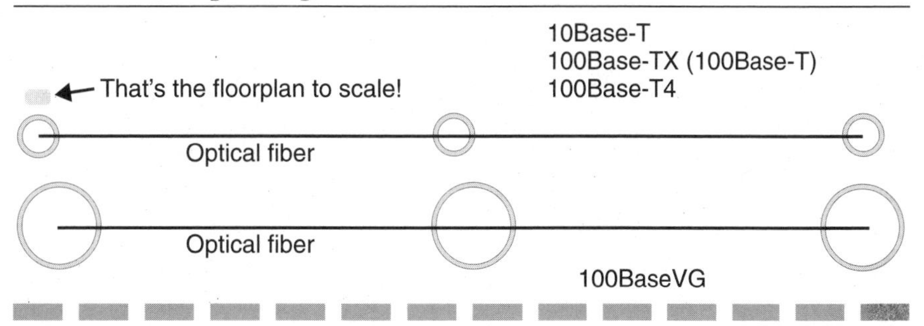

Figure 2.57.
Fast Ethernet wiring coverages with fiber links.

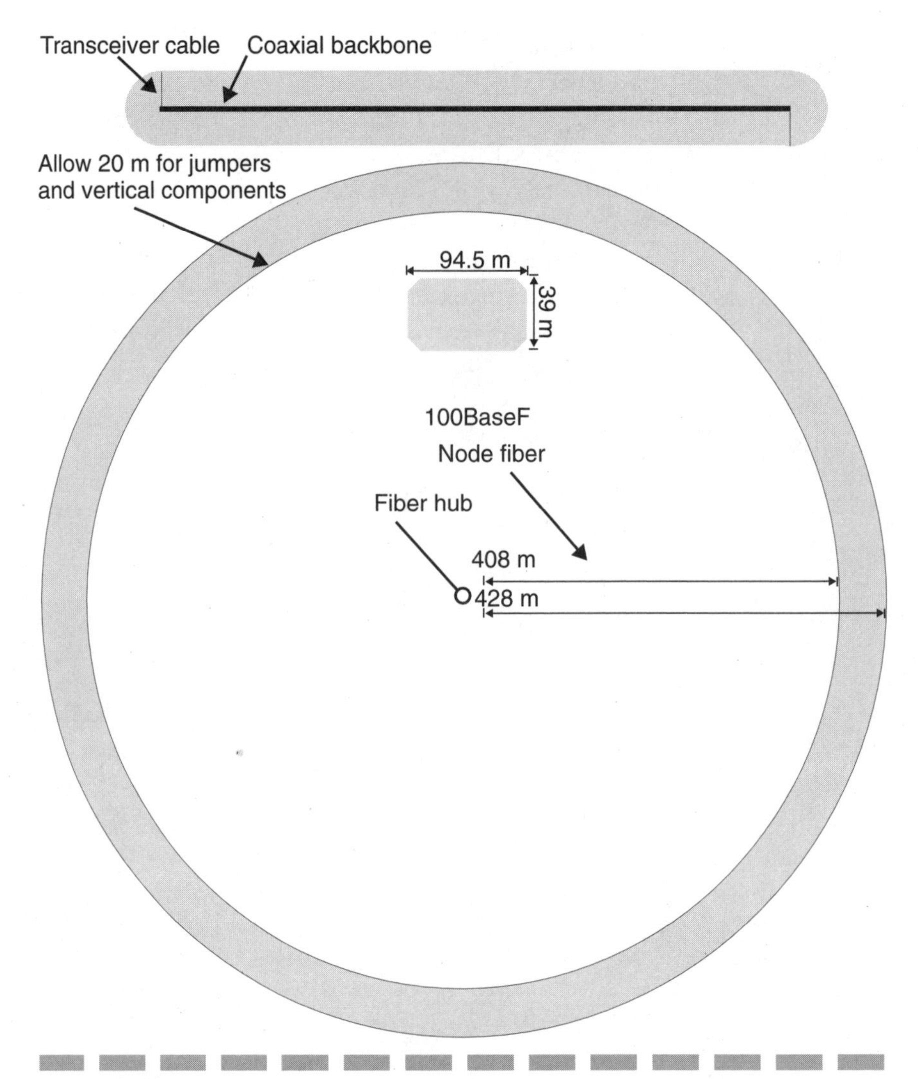

Figure 2.58.
Fast Ethernet optical fiber coverages.

The next illustration shows a cross-section of a floor within an office building. Notice the concrete floor, the floor which is the ceiling above, the steel floor beam, the hanging ceiling, and the hollow plaster walls. When you trace the connection from the PC to the hub, notice that the jumper goes down from the PC to a switch plate, then up from the switch plate over the ceiling and down to a punchdown panel. From there, the connection is patched into the hub. There are both vertical additions to the run and little cords that add to the total overall distance.

In addition, the lobe cable could be measured for a straight run but follow corridors, weave through duct work, and have a few extra loops at each end. Consider all these pieces in Figure 2.59. It is also useful to document all these connecting points, because every one is a potential location for transmission problems. So much of *Fast Ethernet Implementation and Migration Solutions* is about premise wiring because the signaling speeds of Fast Ethernet, which are from two to 10 times faster than Token-Ring and Ethernet, push the limits of the wiring infrastucture.

Interoperability

Fast Ethernet is consistent with the IEEE 802.3 standard for the physical interconnection of nodes on a network. Packet formats conform to the 802.2 link protocols. Duplexed Ethernet creates two unidirectional channels for communication, but is still bound by collision detection on the

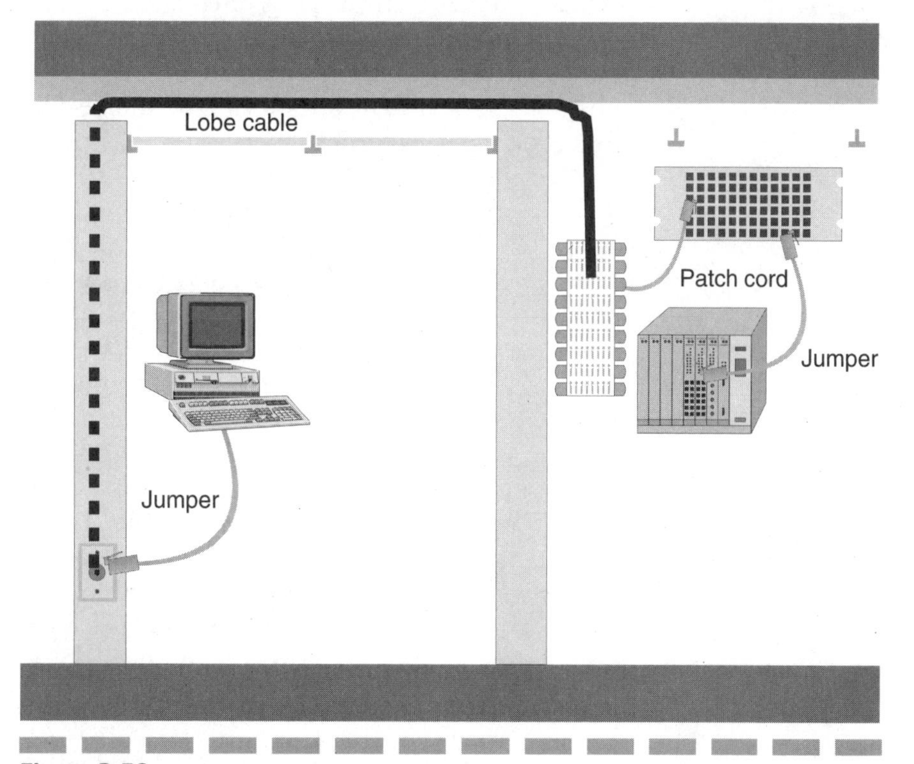

Figure 2.59.
Include vertical components and both jumper and patch cable components to the overall lobe wire length.

shared backbones. Switched Ethernet allows only two nodes on a segment, and it arbitrates transmission so that collisions never occur and there is no need to even check for them. However, because low-end switched hubs concentrate multiple streams into a single backbone inside the device, collision detection is still necessary in those devices. All Fast Ethernet systems conform to specification, with two important exceptions. 100BaseVG-AnyLAN uses four pairs of wires and quartet signaling to enable what is termed a *demand priority* access method, and all port-switching technology disables (or bypasses any need for) the collision mechanism. This includes 3Com's Priority Access Control Enabled (PACE), which is a port-switching protocol for allocation of bandwidth.

Despite apparent similarities among Ethernet family members, considerable incompatibility exists. Ethernet 1.0 is a receive-based collision-detection version, Ethernet 1.1 is a rare refinement of signal quality assurance (SQA), whereas both Ethernet 2.0 and IEEE 802.3 apply transmit-based detection that relies on the collision interference signal. Wireless and switched transmissions often create a collision avoidance mechanism, or simply disable collision detection and control transmission permissions with a different algorithm.

IEEE 802.3u, the 100-Mbps version of 10Base-T, is exactly like 10Base-T except that every Ethernet event occurs 10 times faster. Note that some hubs provide both 10Base-T and 100Base-T support for interoperability with a media and signaling speed bridge built into the hardware. 100BaseVG is designed to interoperate with 10Base-T, 10Base-FX, and 10Base5, but it is not easily bridged to 100Base-T. Because no vendor has provided a bridge with dual support, you will have to connect 100Base-T with 100BaseVG through a common 10Base-T port, or construct a higher-speed gateway to translate the media and protocol. A gateway is nothing more than a PC with 100Base-T and 100BaseVG adapters with routing software. Windows 95, Windows NT, NetWare, and most versions of UNIX handle this well. 100Base-FX at least has become common and very uniform. It should interoperate without problems.

These variants are incompatible with each other, and have a tendency to cause interconnectivity problems when mixed without bridging or routing firewalls. The initial IEEE specification 802.3 adoption was delayed a year because IBM proposed Token-Ring as an alternative and brought a lot of pressure to bear on the IEEE. As a result, the IEEE split the 802 committee into separate subcommittees to define a standard for the logical link (802.2), Ethernet (802.3), token bus (802.4), token ring (802.5), and the Metropolitan Area Network (802.6). DEC, Intel, and Xerox released an upgraded Ethernet 2.0 before the 802.3 formulation was finalized. Compatibility should be viewed in these terms as well, but on a sliding scale. Now we have 802.3u and 802.12 issues too.

100Base

The three variants are: 100Base-T with a wiring infrastructure defined by the 100Base-TX two-pair wire specification; 100Base-T4, which is really 100Base-T with a four-pair wiring system and a different electrical signaling basis, as well as a different bit-encoding method; and 100Base-FL, sometimes refered to as 100Base-FX, which is still 100Base-T but transmitted on fiber with light to carry the signal. 100Base-TX and 100Base-FX uses 4B/5B and MLT-3 for encoding. This is the same signalling method used for FDDI, but the packet format is still Ethernet, and is very different from the FDDI packet format. Note that the fiber signal speed is actually 125 Mhz to support the encoding. 100Base-T4 uses 8B/6T and NRZ for bit encoding. The wire signal speed is 33 MHz on each pair, for a total of 133 MHz over the four pairs. Encoding reduces overall packet throughput to 100 Mbps for all these Fast Ethernet variants. Data throughput is less because of packet overheads, timing issues, and management.

100BaseVG

There is a single definition for 100BaseVG. I think that it really is different from Ethernet because of the signal, protocol, and traffic management features, but at least it was designed to interoperate with 10Base-T and resolve some of the saturation bottlenecks common in a server-based network. The substantial differences include a different traffic management protocol, its innate ability to handle large frame Token-Ring packets (with VG-AnyLAN), and the ability to prioritize traffic in much the same way as SNA. The 100BaseVG wiring infrastructure is the same as defined for 100Base-T4. However, 100BaseVG is electrically incompatible with 100Base-T4 and it also uses a different bit-encoding method. 100BaseVG uses 5B/6B and NRZ for encoding. The wire signal speed is 30 MHz on each pair, for a total of 120 MHz over the four pairs. The encoding reduces data throughput to 100 Mbps for this Fast Ethernet variant.

Duplexing

Duplexing at 10 Mbps or 100 Mbps is really a system to microsegment a LAN into a two-channel architecture permitting simultaneous transmit and receive. Although in principle this should double available bandwidth and hence actual throughput, duplexed 10 Mbps really provides about 14 to 16 Mbps throughput in a Windows NT or OS/2 network where systems can take advantage of CPU multitasking. Duplexed 100 Mbps pushes the limits of most network devices to generate more than 40 Mbps of real-world traffic on one channel. This architecture is useful for aggregating clients on one channel and servers on another, and is an effective migration technology from Token-Ring networks with long latencies or 10Base-T networks experiencing excessive collisions.

Although the basic problems of Ethernet are still there in a duplexed infrastructure, the migration requires substituting only new NICs and hubs on a one-for-one basis, and you can add usable bandwidth. This is useful when user PCs and workstations sustain simultaneous logical network connections, as through a modem, SLIP, or PPP Internet connection, or through multitasking.

Switching

Switching is the ultimate in microsegmentation and traffic filtering with a routing protocol. Various vendor products provide different degrees of network microsegmentation and logical rewiring of LANs with virtual connectivity. When switches are just routers, switching trades bandwidth for increased end-to-end transmission delays. When switches are based in ASIC hardware, they provide an effective method to increase bandwidth, with few pitfalls other than increased hardware costs and network management headache (because of the primitive tools for traffic management on switches, and the lack of comformity to SNMPv2 for most Fast Ethernet switching products). *Reduced instruction set chips* (RISC) are general-purpose computer chips, and represent an older and slower generation of Ethernet switching processors. They do not perform as well, because of latencies equivalent to those of bridges and routers. The next section explains the tradeoffs with switching bandwidth for latency, and how this is a fundamental limitation in Ethernet technology and Fast Ethernet migration strategies.

Bandwidth versus Latency

All versions of Fast Ethernet represent effective migration strategies when upgrading from ARCNET, Ethernet, or Token-Ring single-segment local area networks. Each LAN will perform better after the upgrade; Fast Ethernet provides more bandwidth and faster delivery times. You are unlikely to see 10 times improvement; two- or threefold improvements are realistic. Once the network bandwidth limitations are resolved with Fast Ethernet, most workstations and servers are then limited by disk or CPU bottlenecks. You might have solved one limitation to create a new active bottleneck, albeit at higher throughput levels. 100Base-T and 100BaseVG as a direct substitution for a LAN is always an improvement. Duplex is a boon for a network with unbalanced or uneven traffic loading characteristics, such as client/server. Switching works when bandwidth is at a premium and delays in delivery are irrelevant. The decision for switching is more complex when timing is important, or when you have networks of LANs. Clearly, selection of switched-10, switched-100, duplexed, 100Base-T, or 100BaseVG will matter based on your environment and application mix.

Anticipated performance improvements from migrating to Fast Ethernet are unlikely in complex internetworks supporting traffic between individual LAN segments unless the entire infrastructure is upgraded uniformly, because of the multiple delays (and relative length of these delays) at existing intermediate nodes and the added latencies of new intermediate nodes. When you do improve performance, you are unlikely to see overall improvements greater than 200 percent. In fact, you might even degrade performance for applications distributed over networks that are data-bound, because of the tradeoffs between bandwidths for switched latencies. There is no simple answer as to when Fast Ethernet works or doesn't work. It just isn't that simple. Figure 2.60 suggests realizable performance improvements:

Optimization of networks (inclusive of WANs, LANs, internetworks, and enterprise components) must begin as a fundamental review of the network infrastructure. Although statistical modeling is essential for forecasting loads and response times under different network applications, protocols, designs, and technologies, the relationships between bandwidth and latency must be understood. Microsegmentation (the techniques of subnetting, LAN segmentation, and switching) for bandwidth enhancement is only effective to solve network overloads and bottlenecks, not performance delays; in fact, segmentation adds delay.

	10Base-T			100Base-T			FDDI		
	STD	Duplex	Switch	STD	Duplex	Switch	STD	Duplex	Switch
10Base-T	Parity	1.4	4	4	5	12	5		12

Figure 2.60.
Performance enhancement matrix with Fast Ethernet.

While faster protocols (such as Fast Ethernet, of course) and microsegmentation (switched connections) increase network bandwidth, these techniques are most applicable when network performance is compromised by extreme overloading and collision saturation. The disadvantage of microsegmentation is that it also increases the latency of internetwork transmissions due to service times at intermediate nodes. In simpler words, hubs, repeaters, bridges, routers, switches, and all-in-one enterprise devices increase the end-to-end transmission times. Network performance optimization is an issue of fundamental design. It is not a post-operative facade. No amount of reengineering with Fast Ethernet or switches can redress design problems.

Network and application designers who prophesy performance enhancements through the wholesale application of Fast Ethernet will instead degrade overall internetwork performance. They trade asynchronous LAN bandwidth for increased end-to-end and point-to-point packet transmission latencies in complex networks. *Latency* is typically defined as the service time required to process a packet; this is represented by the processing time for an intermediate node (such as a repeater, bridge, router, switch, or gateway) to actually service that packet rather than delays and overheads actually required under various loads. A more realistic and useful definition of latency includes the end-to-end, point-to-point delivery times including all delays at all intermediate interconnections.

Cutting-edge protocols besides Fast Ethernet include ATM, FDDI, Fibre Channel, SONET, and SMDS. These new technologies typically represent packet switching and virtual networking (or variations on those). Forget the crystal ball; basic protocol specifications and more complex queueing methods can show that the resulting performance effects from high-speed protocols and switching is beneficial only when the core network protocol is unable to sustain the baseline transmission load. Figure 2.61 illustrates bandwidths for common fast protocols and switches.

Networks supporting mixed speeds and hybrid protocols also create serious bottlenecks at intermediate boundary nodes, that is, nodes that provide a transmission speed or protocol transition. This is true because intermediate nodes add service latencies that are at least a magnitude greater than the average LAN protocol end-to-end, point-to-point latencies. The division of a single LAN into two or more subnets is called microsegmentation or network partitioning. Switching is the ultimate example of network microsegmentation where individual paired nodes (that is the two communicating with each other) are configured on-the-fly into virtual subnets for the express duration of that transmission.

The message you should remember is that microsegmentation or network partitioning is only effective when LAN latencies are so lengthy that even with the much longer intermediate nodes latencies, 1) overall end-to-end, point-to-point latency is reduced, and 2) bandwidth is increased without any corresponding increase in end-to-end, point-to-point latencies. Figure 2.62 shows end-to-end latency ranges. The dappled bars show ranges of performance that are based on actual network design and loads, but also on actual hardware performance.

Specifically, as an example, FDDI switching or microsegmentation will not be effective until switching hardware can establish connections or sustain on-demand connections faster than 1.612 ms, the maximum FDDI token rotation time (TRT) under maximum load. However, switches and microsegmentation are effective replacements for overloaded infrastructures when sustained Ethernet, Fast Ethernet, or 16 Mbps Token-Ring loads create latencies that exceed the setup and transmission time for switched connections or exceed the service time for bridged and routed interconnections. Also, note that hybrid 10Base-T/100Base-T

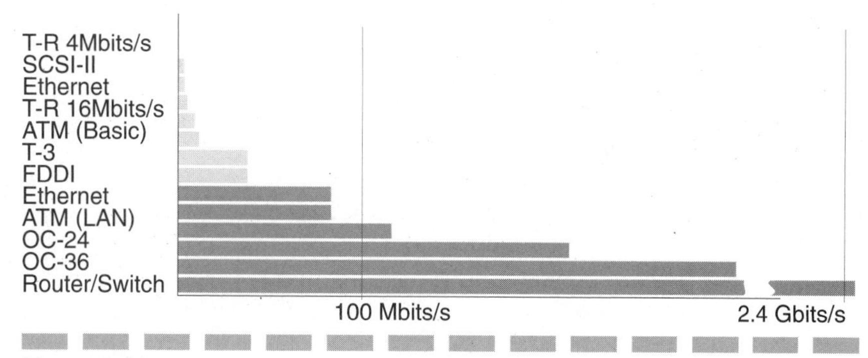

Figure 2.61.
Bandwidths for common network protocols.

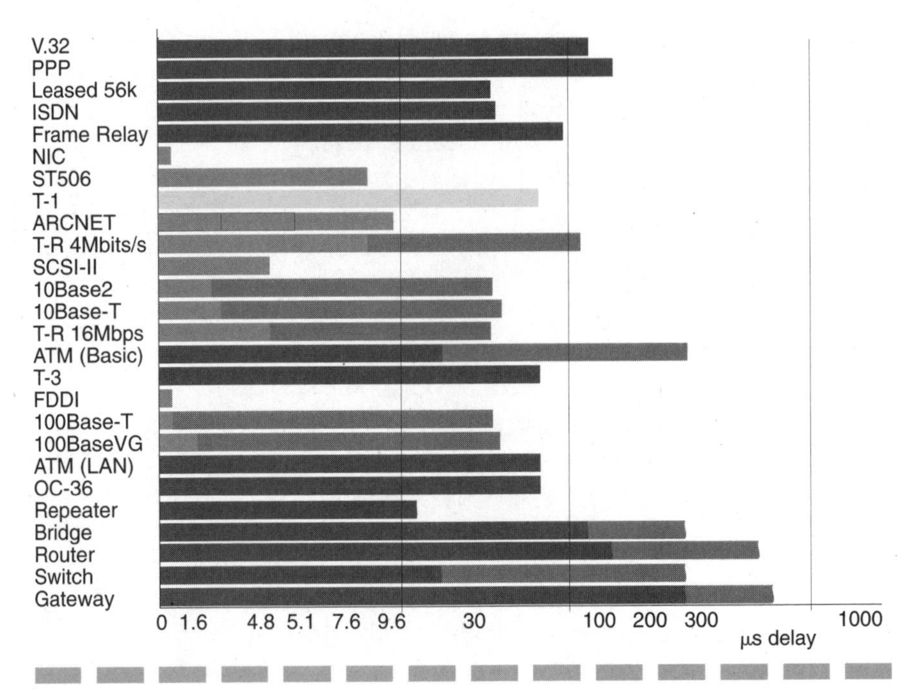

Figure 2.62.
Performance ranges for the end-to-end, point-to-point latencies of some common protocols and intermediate nodes.

is limited not only by the slowest link in the 10Base-T component, but also by the time required by the hybrid boundary hub to bridge packets between the segments running at different speeds, about 80 ms. As a result, transitional Ethernet/Fast Ethernet hybrids for LANs usually do not perform well. The generic example when hybrid migration works well is when you can segment loads between LANs, or you have a disproportionate traffic load from a single node (for example, from a server) that is distributed to many end nodes.

I can offer no generalized designs that work best. Traffic is a complex factor of packet sizes, network node count, and network workload characterization (that is, packet arrival times and variances). That means every network is different. The window of opportunity is also affected by the service and switching times of these intermediate nodes. Intermediate nodes create latencies in the very wide performance range from 28 to 800 ms, although this increases with packet drops, losses, saturation, or bursty overloads.

You need to put a protocol analyzer on your network and capture latency information. For example, if a bridge, router, or switch can establish, transmit, and take down that connection within 100 ms, it becomes a performance *enhancement* to subnet or employ switches on a network with point-to-point service times in excess of that 100 ms, but a performance *degradation* for latencies less than 100 ms. Microsegmentation is also effective to boost bandwidth when latency is not a performance issue for a specific network environment, such as noncritical office support activities, peer-to-peer networking, and utility networks. However, in many cases, client/server application design or overloads at a server or host create delays that overshadow even long network latencies, so much so that network latencies are relatively minor annoyances. This window of opportunity is best illustrated by the graphs in Figure 2.63.

True Effects of Latency

The significant disadvantage is that the intermediate nodes create a queueing delay for cross-traffic. Although most intermediate nodes today can sustain the full bandwidth when appropriately buffered, overall transmission latency suffers because the raw service times for these intermediate nodes is greater than normal LAN transmission latencies—by a magnitude or more. Ethernet, Fast Ethernet, and FDDI preclude simultaneous, overlapping transmissions on each LAN. However, network traffic (from different LANs) can arrive simultaneously at each intermediate node. As a result (remember that the intermediate nodes are dual- or multiported and instead must support simultaneous traffic) the service queueing delay is more substantial than the service time

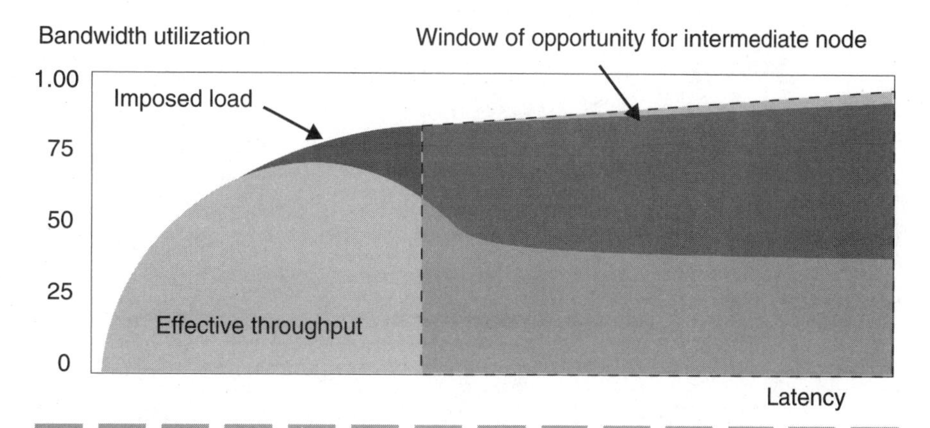

Figure 2.63.
The window of opportunity for microsegmentation is a function of the network workload characteristics and intermediate node service times.

itself, because each queue message must wait for one or more prior messages to be serviced first.

When intermediate nodes are not buffered (as with cut-through switches) in order to provide the fastest service time and ostensibly bypass this bottleneck, simultaneous transmissions are blocked and dropped instead, which is networking's "busy" signal. This creates even greater network service delays because service time-out or non-response retransmission is usually several factors longer than the service times of even the slowest intermediate nodes. Although some hosts, super-servers, or engineering workstations are typically faster and have more CPU power than bridges, routers, or switches, these devices still process software applications in memory, resulting in lost transmissions that are not necessarily noticed by the server, host, or intermediate nodes for many seconds; not milliseconds, or even microseconds.

Performance Fundamentals

Network and application designers should review the fundamental performance characteristics of new "high-speed" network architecture, microsegmentation, or hybrid-protocol infrastructures before embarking upon retrofits and rewiring. Although those Fast Ethernet variants do increase available bandwidth, the tradeoff is usually seen in terms of additional signal transmission latency through any intermediate nodes. Latency is particularly a concern where mixed protocols are translated or encapsulated at a hub, bridge, router, switch, or gateway because unqueued service times are likely to range from 28 to 800 ms. NetWare administrators might recognize the concept of the intermediate node as the "hop." Queued service times typically are multiples of 2 to 3 (that is, two or three prior messages in queue for transmission by the intermediate node) of unqueued service times, while severe performance bottlenecks create queued service delays that include complete service time-outs. This is at least one to two orders of magnitude greater than the transmission latency on any primary LAN. Figure 2.64 illustrates latency times for common network structures and intermediate nodes.

This raw information shows the basis for performance problems with hybrid 10Base-T/100Base-T networks, and the reason why microsegmentation is only a solution for some LAN bottlenecks. On the other hand, all other protocols require either a round-robin token or active hub-signal repeaters (which add no less than 20 ms in latency).

Protocol or device	Bandwidth	Transmission latency: minimum	Transmission latency: mean	Transmission latency: max
4Mbps Token-Ring (250 nodes)	4,000,000 bps	9.6 ms Transmission latency: minimum	38 ms	390 ms
16 Mbps Token-Ring (132 nodes)	16,000,000 bps	7.8 ms Transmission latency: minimum	33 ms	140 ms
10Base5	10,000,000 bps	0.16 ms Transmission latency: minimum	18 ms	unlimited
10Base-T	10,000,000 bps	25 ms Transmission latency: minimum	25 ms	unlimited
100BaseVG-AnyLAN	100,000,000 bps	25 ms	25 ms	58 ms
FDDI	100,000,000 bps	0.32 ms	0.50 ms	1.612 ms
Hub (active)	based on protocol	22 ms	22 to 70 ms	300 ms
Bridge	based on protocol	80 ms	300 ms	unlimited
Router	based on protocol	170 ms	200 to 300 ms	unlimited
Switch	based on protocol	60 ms	300 ms	unlimited
Gateway	hardware-driven	200 ms	600 to 800 ms	unlimited

Figure 2.64.
Bandwidths and latencies for some common protocols and intermediate network devices.

FDDI is not an exception, although its token rotation time (TRT) is limited to 1.612 ms, rather than held above a minimum as provided by IEEE 802.5 Token-Ring specifications. Maximum wait times for Fast Ethernet are driven only by packet size, network node counts, network traffic, and collision effects. Figure 2.65 illustrates latency at a fixed packet size for node counts.

The fastest time in which a hub can retransmit a packet is about 20 ms. While some available-in-the-future cut-through switch is envisioned to provide switching times from 10 to 40 ms, current technology yields the same performance as a network bridge (ranging from 120 to 300 ms, with availability at the higher end). Additionally, because the cut-through switch cannot buffer packets and is thus a single-server with no queue in spite of its inherent bidirectional connectivity (for two or more nodes or subnets), actual performance under realistic network load is worse than a bridge. The modified cut-through switch (that is, a switch with a buffer) provides the same disadvantages of the cut-through switch and all the performance characteristics of the router, too.

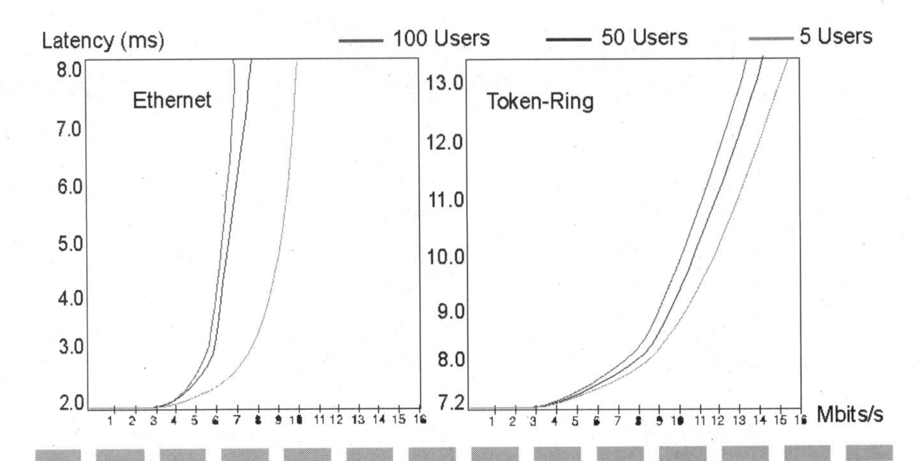

Figure 2.65.
Latency as a function of node count. As with other Fast Ethernet times, latency on Fast Ethernet is 10 times less on simple LANs than on complex ones.

There is also the issue of circuit blocking with these matrix switches, whether cut-through or buffered, because switches prevent establishment of multiple connections to the same node (important when a node is a server, host, or intermediate node). In other words, that is a "busy" signal. Newer switches do have a two-tiered architecture so that all traffic is dumped into a buffered backbone. Realize that traffic bound for a busy server gets queued up and is delayed in that buffer. In addition, there are other significant delay points.

Although this information is fundamental to basic network design, and most of you know it fully, you might not have analyzed the performance aspects. It might seem immaterial, or basically you saw no feasible alternative to these segmented network designs. Also, latency previously might not have mattered at all, but it does now, and will very dramatically in the future. However, even LANs of 40 users (mean value) are driven by the technology required by 5000-user internetworks, downsizing, upscaling, client-server applications, host connectivity, multimedia, videoconferencing, and primarily by just more of the same—usually a lot more of the same.

As a result, network performance engineering is becoming a critical issue for any organization with data networking. I want to reinforce the concept that any network performance engineering is firmly rooted in fundamental design issues and cannot be a facade built with retrofits, workarounds, and high-speed linkages. The key parameters include system and network design, model techniques and the model spaces, workload characterization, and the fundamental physical limitations in the processing and network infrastructure.

All network protocols apply a signal transmission speed that ranges from 0.58 to 0.88 light speed (3×10^{10}); any transmission signal speed differences are really immaterial. Something else—specifically the delays at servers, clients, and intermediate nodes, and also packet errors and losses—creates performance problems. The network is really a pipe with a volume and a pressure. Figure 2.66 shows this.

The volume is the bandwidth (in conjunction with transmission speed), but the pressure is really two things. Part of the pressure is the signal transmission speed, and the other part is the turbulence of packet arrival times. Turbulence is defined as the burstiness or peak-load characteristics of the traffic volume. (This is represented mathematically by the volume variance and mode and the volume arrival time distribution.) This turbulence is a function of multiple node access to the singular "shared media" and the workload characteristics for each node. Every network is basically a shared infrastructure. Although it is possible to create dedicated connections between nodes with virtual LANs, switches, and other new technology, fundamentally, nonetheless, the network is predicated by the sharing of processing, intermediate nodes, or point-to-point transmission bandwidth. Figure 2.67 shows how turbulence from bursty traffic loads breaks up the signal flow, wastes available bandwidth, and delays delivery. Although signal delivery on an individual

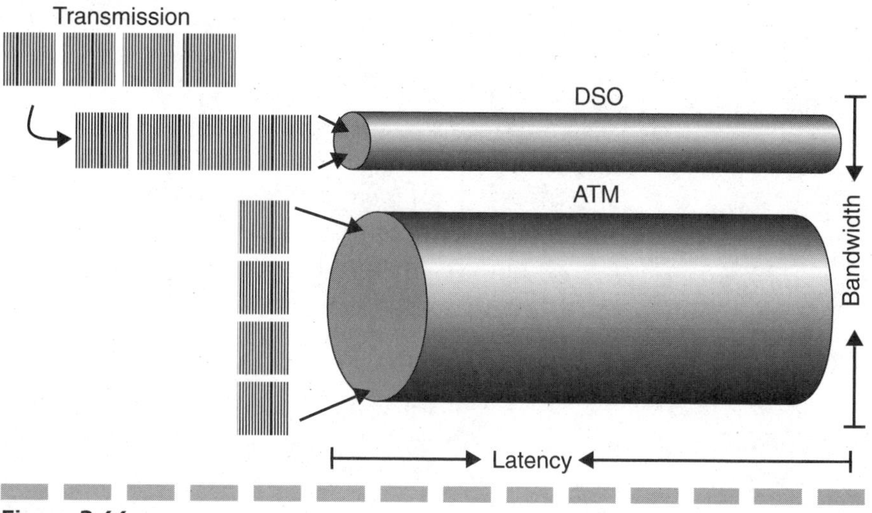

Figure 2.66.
The contrast between bandwidth and latency.

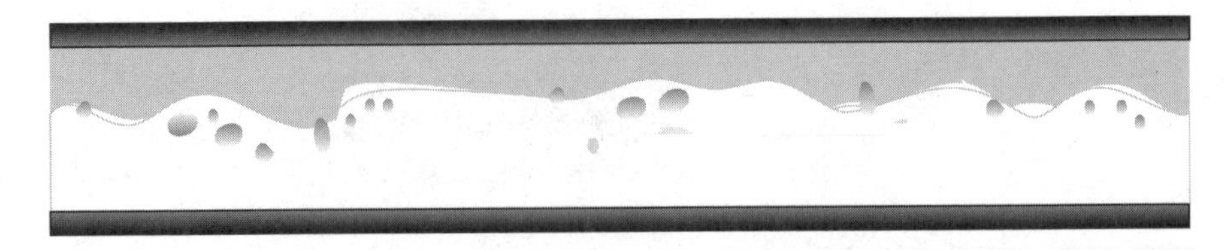

Figure 2.67.
Turbulence is a function of message arrival times and network workload characteristics.

wire (that is, the metaphorical pipe) is very precise, point-to-point delivery includes many transitions that create turbulence.

Although network load is a function of mean packet sizes driven by the network node counts, the turbulence is a factor of the workload characteristics. If traffic is bursty, the infrastructure will saturate more readily at intermediate nodes because the service times are much longer there. In fact, queueing analysis easily shows that bursts create more latency than large packets or fully populated networks; hence you can equate the network flow turbulence to the inherently bursty network traffic. (Real-world emulation of internetworks with bursty traffic is very difficult, because it is exceedingly difficult to replicate it in a test environment; this is the reason most router and particularly Ethernet and Fast Ethernet switches' measurements are based on trivial block/file transfers and provided in terms of packet/s forwarded.) The traffic bursts create queues at the intermediate nodes with constraining service times. The mechanism for the queue (such as FIFO, LIFO, etc.) is immaterial; the mean total service and waiting time will be exactly the same unless there is priority service and priorities are implemented correctly.

As a result, "slow" performance is often a function of delayed response times, which in turn is the end-to-end, point-to-point performance latency. This is typically equated with bandwidth limitations—incorrectly, I might add—because it is the only parameter easily controlled through retrofits, upgrades, switches, or adding more channels through products from vendors. As explained previously, when Ethernet reaches critical bandwidth saturation, excessive collisions create total bandwidth saturation. Similarly, when FDDI's token-based protocols are stressed by increased bandwidth utilization, the TRT increases exponentially. This, of course, leads to slower message delivery times, which is longer latency.

The perception is that latency is fixed by the speed of light, by the workload characteristics, and by the bandwidth of the carrier channel. Obviously, increases to the channel bandwidth generally improve performance when latency does not increase in concert. For example, shift from 1200 bps to 56 Kbps, go from first-generation LAN technologies to high-speed Ethernet or FDDI, and migrate from T-1 to OC-192 to improve performance. In the same way, shift from one network to two parallel networks and split the loads between them. This is fundamentally less costly, and is a common procedure. In fact, the most common fix for slow LANs is microsegmentation.

Although intelligent partitioning and redistribution of the network nodes to the new segments is a primary requirement for success with this approach, there are other indirect factors. Specifically, cross-traffic among the new LANs reduces total available bandwidth below than raw bandwidth. Some protocol substitutions yield substantial benefits. For example, a shift from 10-Mbps Ethernet to Fast Ethernet truly represents a bandwidth increase of 1000 percent, and also a substantial reduction in latency, because the minimum signal propagation time is shorter, and all event times are shortened by a factor of 10. However, there are often secondary results that must be investigated. Consider that a fully populated local area network at the higher speed might be able to support a maximum of 132, 72, or as few as 20 nodes (depending upon the physical wiring hub). Under this scenario, the original LAN will need to be partitioned into two or more subnetworks through one or more intermediate nodes. When a ring (or even a bus or star) topology is thus segmented, an intermediate node is required to interconnect the partitioned structures, as Figure 2.68 shows.

The Ethernet protocol, which underpins 50 percent of all data networks, creates a different performance issue when service arrivals exceed mean service time during each successive collision interval (51.2 ms at 10 Mbps

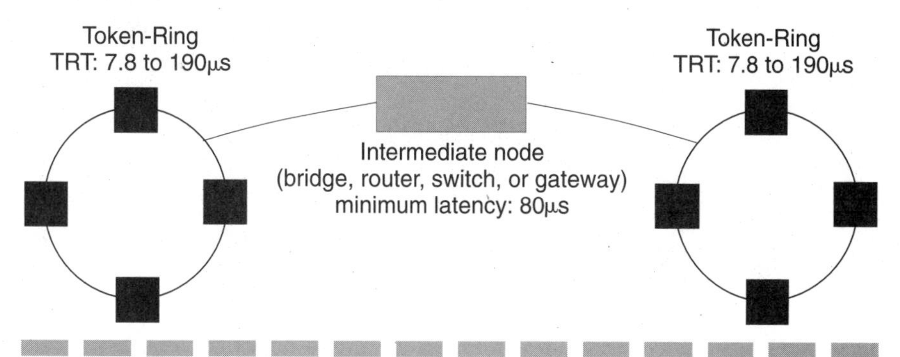

Figure 2.68.
Required ring partitioning to support the upgrade from 4-Mbps to 16-Mbps Token-Ring.

and 5.12 ms at 100 Mbps). If the arrival rate is sustained above the service time (the basic equation for poor network performance) network throughput drops as collisions saturate bandwidth. Latency also increases dramatically, due to packet retransmissions of at least the packets causing each collision. Ethernet utilization reaches a point where collisions increase to the detriment of data throughput; formally, this occurs when the mean arrival rate exceeds 1.63 to 1.78 packets (a function of packet size). In other words, this means that more than one packet arrives during the formal collision interval and creates a collision. This saturation is best illustrated by Figure 2.2, back near the beginning of this chapter.

Microsegmentation is a particularly effective solution for Ethernet bottlenecks caused by saturation, because the network latency under collision saturation becomes excessive, easily surpassing intermediate node latencies. A single collision costs a mean 25.6 ms (uniform mean on range 0 to 51.2 ms for first collision backoff) and increases exponentially with succeeding sequential collisions due to the standard implementation of the binomial logarithmic backoff mechanism for resolving collisions. Because an arrival rate of merely 2.01 packets per second will sustain a mean latency delay of 225 ms, bridging provides a viable solution to the saturated 10-Mbps Ethernets. You are basically trading 225 ms in collision time for 80 ms in bridge delay, which is a sensible tradeoff. However, subnet partitioning at a server yields much better performance, as Figure 2.69 illustrates, because there might be no service delay between clients and the shared server if internetwork traffic (and the need for the gateway) can be totally eliminated.

While these microsegmented designs are sensible for 10Base5, the comparison is more complicated for 100Base-T, 100BaseVG, and hybrid networks. While complete 100Base-T substitution improves an overloaded 10Base5 or 10Base-T LAN because bandwidth is increased and latency is decreased concurrently, internetworks with intermediate nodes or mixed applications of 10Base-T and 100Base-T will include the hub latency of at least 20 ms or bridge and routing latencies of at least 80 ms. In fact, The European Network Laboratories in 1995 benchmarked most switch latencies for real products within the range of 71 to 90 ms for the minimum sized 64-byte Ethernet packets. In other words, do not migrate to a switched infrastructure unless the performance delays are greater than the switching times. Hence, only under extreme performance loads where latency is excessive is this migration advisable.

When network performance averages 20 to 200 ms, the migration decision is more complex, because the subnetted partitions often exhibit var-

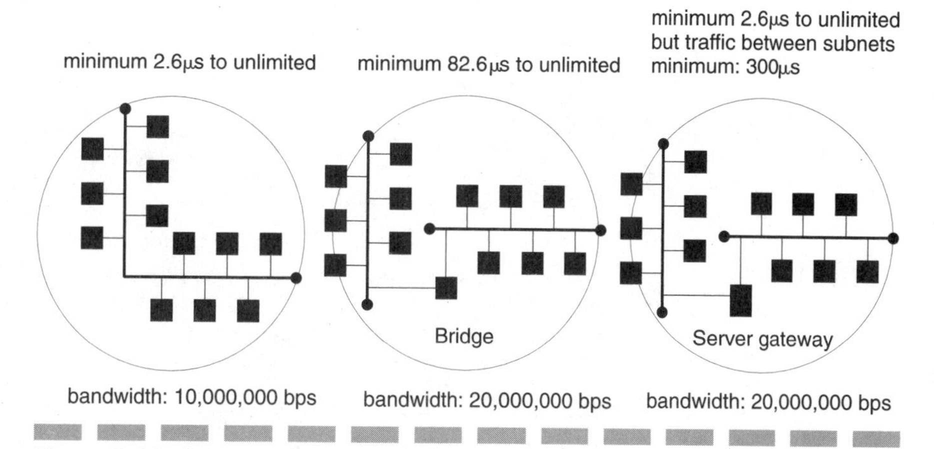

Figure 2.69.
Maximum raw bandwidth for a saturated Ethernet subnetted at server and by a bridge. Your utilization will vary based on internetwork traffic loading. The signal latency is depicted by the circles.

ious latencies of their own. Note that the latencies are not strictly additive (either sequential or synchronous), merely cumulative in a complex bidirectional chain of queues. Although the latencies on the subnets can be substantial, usually intelligent partitioning results in subnets with low loads. Nevertheless, the new internetwork now represents a series of bidirectional service queues at the intermediate node and service queues to access each subnet, all of which can create new performance delays, as shown by Figure 2.70. Note that the intermediate node connects to two or more subnetworks. That bridge could be a network server with multiple network ports or a multiport adapter card.

The ultimate microsegmentation technique is virtual switching, whereby each pair of communication nodes are switched to a unique circuit. This provides full bandwidth for the conversation and no latency beyond the switching time. As a result, the available bandwidth is a multiple of the number of switches available. If all nodes have full-switching access, the bandwidth allocation is full bandwidth as limited only by the protocol to each and every node. This is the promise by switch vendors. However, correct infrastructure analysis requires that the switch latency be subtracted from available bandwidth. Because it only takes 16 ms to transmit the longest possible Ethernet frame and 1.6 ms to transmit the longest possible Fast Ethernet frame (data packet with preamble, synchronization, and trailing delay overhead), available bandwidth is reduced to 5,000,000 bps and 50,000,000 bps with a very fast (and as yet

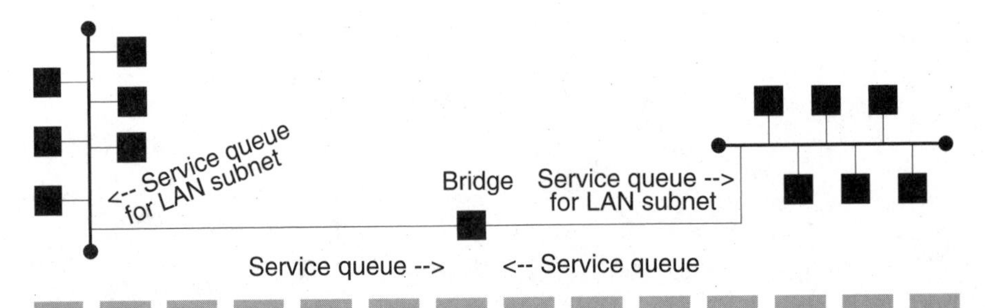

Figure 2.70.
Partitioning an overloaded LAN into subnets could also create a series of new performance bottlenecks at each service queue, bottlenecks that exceed the average latency of the original LAN.

unachieved) switching time of 16 ms. This drops to 1,280,000 bps (12,800,000 bps at 100 Mbps) at switching times of 80 ms, and to a very slow 533,333 bps (5,333,330 bps at 100 Mbps) at the realistic switching speeds of 300 ms.

When Ethernet or Fast Ethernet is routed over either FDDI, ATM, frame relay, or SMDS, and the base protocol is not native, factor in a protocol translation or encapsulation time. Because LAN protocols support maximum packets sizes that range from 637 to 18,432 bytes, expect translation or encapsulation times to range from 70 to 500 ms. Cisco quotes encapsulation times of 20 ms, but I haven't corroborated this. The decoding is needed at the other end, too. Because most network transmissions are directed to a single node and are not broadcasts, more bandwidth is lost because nodes are blocked from transmission or packets are queued because the service mechanism is busy. Again, simple mathematical comparison shows that a range of opportunity exists for the new technology, but it is no panacea for all conditions. When queues are introduced at the intermediate nodes, the comparison only gets worse, and the window for opportunity is narrowed more.

The wholesale application of cutting-edge protocols and new technologies will degrade overall internetwork performance when latency is not considered. Service delays are important for successful implementations of client/server applications, LAN downsizing of mainframe processes, and remote-site processing with multiple hops. Because this describes all but the most primitive network environment, point-to-point, end-to-end transmission latency is a critical design factor in a network. Bandwidth cannot substitute for poor latencies, but you can increase bandwidth in the trade for shorter latencies. You can trade some protocols for others to minimize latencies in the LAN environment, but the speed of light fundamentally dictates WAN transmission delays. However, you can minimize those latencies to a theoretical minimum by collapsing the network topology to eliminate as many intermediate nodes as possible. Even with substantial performance improvements in intermediate node hardware and software, end-to-end latency through the intermediate nodes is still several orders of magnitude greater than that of the simple LAN infrastructure. Even with switching, fast routers, and duplexed transmissions, the baseline internetwork performance is slower than LAN wire speed. What many designers forget with this sudden relative network bandwidth performance improvement is that the complex internetwork always has this substantial packet transmission latency at intermediate nodes.

Therefore, applications designed and tested on LANs with 2- to 32-ms packet transmission latencies must operate within an infrastructure with an unidirectional base latency ranging from 300 to 900 ms. Internetworks with multiple intermediate nodes (that is, "hops") yield one-way packet transmission latencies of seconds. Additional microsegmentation and switching is not a solution for reducing this latency, although collapsing the network backbone will address some of it. Other performance issues represented by excessive latency might require a fundamental redesign of client/server applications or additional code to automatically handle these transmission delays.

For example, migration from Ethernet to Fast Ethernet will generate a transmission latency savings of 38 ms (from 40-ms transmission times to 2 ms). That is a transmission speed improvement of 2000 percent. However, 38 ms is small in comparison to server processing times. Consider that a database server averages 600 ms to recognize a data request, and another 800 ms to fulfill it. If you add the numbers, the initial overall transaction time is 1480 ms, and this is reduced with Fast Ethernet to 1404 ms. The improvement within this context is only 5 percent. That difference might not matter at all. Do you double your networking costs for a 5-percent improvement? Do you promote Fast Ethernet as a solution to user response time bottlenecks in that case? You want to put your migration strategy in context with an overall performance goal.

Performance Considerations

Migrating to Fast Ethernet is a no-brainer for LANs and a moderate modeling task for most interconnected networks. Complex enterprise networks and client/server infrastructures represent complex data highways that might not benefit from an overall migration to Fast Ethernet. Although parts of the network will benefit from the faster transmission clock speeds, the overall latencies are difficult to model with assurance. In other words, performance for local traffic will benefit with a migration to Fast Ethernet, but increased latencies for intermediate nodes and switches will affect users accessing services from remote segments. You have to understand the hardware limitations and the design decisions.

Switching and duplexed network technologies provide greater bandwidths than the basic Ethernet or Fast Ethernet protocols. However, understand the performance characteristics of the switches and the ramifications on network performance. Some saturated networks benefit a little from switched infrastructures, but other bottlenecks create new delays and performance problems. The switch vendors dangle the promise of 100 Mbps to 5 Gbps of bandwidth, and to a large degree the vendors deliver. Contrast this bandwidth to the ability of hosts and servers to generate more than 20 to 30 Mbps of traffic, and you should understand that network performance is much more than the channel and the size of the network switch bandwidth.

Most organizations take steps to migrate from Ethernet to switched Fast Ethernet. There is a good reason for that. All the performance benchmarks that I have seen tend to overstate the benefits of Fast Ethernet migration, because operational networks are more complex than simple network test beds created in a lab, and the typical workflow applications are more complex and profound than the usable benchmarks. TPC-A, TPC-B, NetBench, and Perform3 come to mind. Therefore, consider migrating servers and other nodes bound by network speeds to a switched 10-Mbps Ethernet or shared-media 100-Mbps environment.

Network bottlenecks are not so obvious that a sudden and complete migration from Ethernet to Fast Ethernet is the best idea, even when organizational finances make such events possible. I prefer the subtle strategy of migrating pieces to a faster protocol or switched environment on an as-needed basis. Prove to yourself that the incremental bandwidth increases actually improve performance. Switched 10 typically requires no recabling whatsoever, just a substitution of the standard 10Base-T hub for a switch. Because servers and other high-bandwidth users are often in the computer lab or wiring closet, migration to mixed 10/100-Mbps environment often means a new NIC, a new hub, and Category 5 UTP from those high-bandwidth users to the hub. This may be an easy job of changing a few jumpers, or just a little bit harder if you need a few cables pulled through the ceiling, under the floor, or through conduit.

When you can demonstrate that these nodes are still impeded by network bandwidth, make the next step. Pull new cables, buy new NICs and hubs, and migrate to 100Base Ethernet or switched 100-Mbps Ethernet. As you increase the overall network traffic load, you are likely to increase the backbone traffic levels and shift the bottlenecks from servers to the backbone. At that point, you can reassess backbone microsegmentation and whether a complete migration to switched architecture is warranted.

Although the value for used Ethernet equipment is very low, almost to the point of making it impractical to recycle or resell components, consider that a slow and analyzed migration from Ethernet to Fast Ethernet is likely to provide more beneficial performance improvements, fewer setbacks, and be less disruptive than a massive redeployment and a switchover during a single weekend. Furthermore, if you define possible end positions for Fast Ethernet and migrate with a slow and steady progression, you can pass equipment down the food chain from servers to workgroups as you install more sophisticated equipment at each new bottleneck as it is identified. So, while you may be phasing out the Ethernet components that are no longer expensive or valuable, the various Fast Ethernet variants are likely to be valuable elsewhere on the network.

You can explore the many Ethernet variants, network wiring options, and their performance effects with information from a Windows utility database called NETSpecs. As shareware, it is available on CompuServe and the Internet. The database includes information about many common protocols, including Ethernet, Token-Ring, and FDDI, as well as telephony and WAN connectivity. Different screens provide information about electrical or optical cable properties, site preparation and installation instructions, and critical wiring limitations. This database also includes multimedia in the form of pictures of topologies, slices of cable materials and cross sections, sound recordings about the protocols, and AVI footage showing installation and debugging. Figure 2.71 shows a NETSpecs in action.

Financial Comparisons

Network migration is really a factor on finances. If your organization cannot pay for new cabling, new NICs, and new hubs, better performance with faster protocols and new Ethernet variants is a dream. Consider: Wiring and hardware components for a new 10Base-T network are

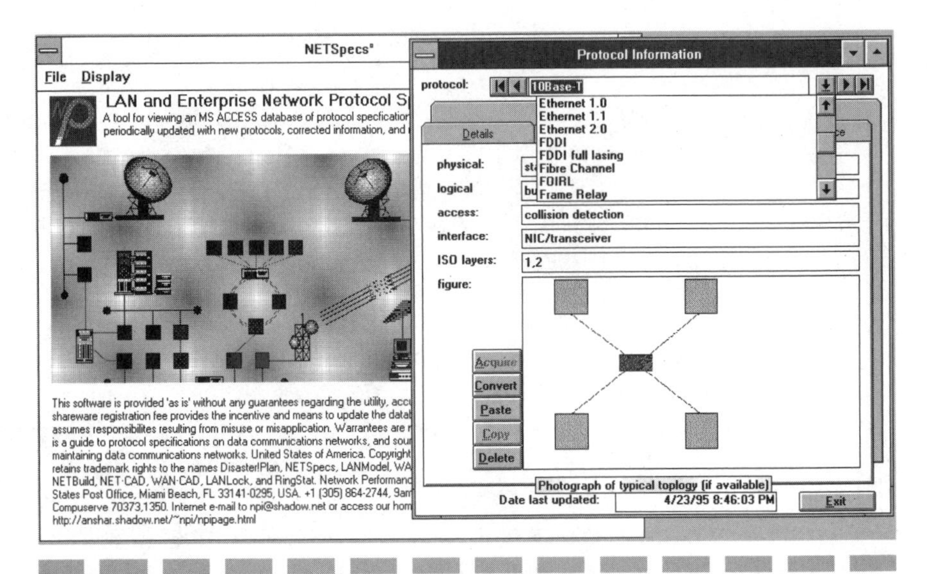

Figure 2.71.
NetSpecs 3.x multimedia database. (*Network Performance Institute*)

about $250 per port, and $200 of that is just in the wiring. Unmanaged hubs cost about $20 per port, and quality NICs are available for as low as $15 each. Add about $200 per hub for management tools (and I advise you to have hub-based management, because the initial cost is cheaper than just one event and even one minute of network downtime). Various consulting groups estimate downtime costs $1000 per minute on small LANs and $100,000 on large telemarketing networks.

There is almost no salvageable value in the preexisting network that you strip for the migration. Contrast this with costs of $400 to $750 per port for either 10Base-T duplexed or switched connections, $200 for 100Base-T, and $400 for 100Base-T4 (although that soon might be on parity with 100Base-T) or 100BaseVG. Duplexed, switched, and/or switched and duplexed 100Base-T costs about $1200 per port, on par with local ATM. All these figures include $200 in the wiring. Wiring costs may be more for 100Base-T4 or 100BaseVG because of the extra pairs, and also because a 100BaseVG link can be as long as 200 m.

Those are the costs for migration. You migrate to Fast Ethernet because you cannot tolerate the ongoing costs of poor performance under Ethernet or similar network protocols. If downtime costs from $1000 to $100,000 per minute, consider what marginal network performance costs the organization. For example, consider that poor network performance creates a 10-percent performance loss. Unlike downtime, which is hopefully limited to a few minutes here and there, poor performance typically persists for hours through the morning and midafternoon. Although chronic network performance problems usually get worked around so that users adjust what they do and when, there is still a financial loss related to it. For example, consider this performance chart in Figure 2.72.

This graph shows 12 hours of network pain. When you compute financial losses, you will find that 720 minutes are adversely affected. If you use the lower $1000/min. cost, chronic network slowdowns cost $72,000 on a daily basis. The higher figure yields a $7.2 million loss on a daily basis. As you can easily see, this network inefficiency is a potent financial incentive for fixing the problems. Payback may happen in less than a day. In reality, the decision to correct network performance problems is not seen as so clear-cut and so obvious. Nevertheless, this type of analysis is the logical tool to formulate migration paths.

The difficult-to-determine numbers in this type of analysis are network downtime cost, a corresponding performance loss percentage for suboptimal network performance, and the terminal network performance after Fast Ethernet migration. The prior comparisons assumed that network performance would be perfect after the network migration. I would not let such an assumption stand unchallenged, and I would not assume a Fast Ethernet migration would address every chronic network performance problem. Inasmuch as 100-percent solutions are

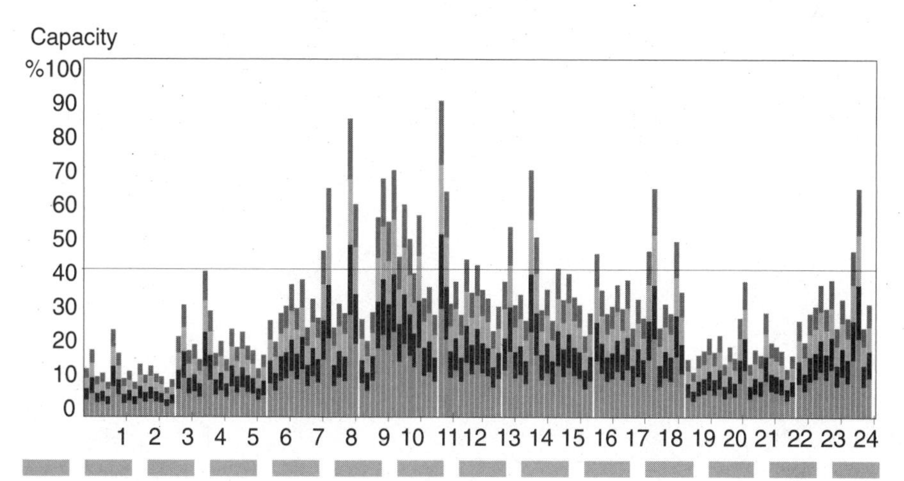

Figure 2.72.
Poor performance persists on a daily basis.

the stuff of financial decision-making, be realistic and try to plan realistic expectations of network redesign. Improvement by only 25 percent suggests a daily financial benefit of $18,000 to $1,800,000. Numbers like those should be enough to cost-justify most network improvement projects, but also be realistic enough not to oversell the benefits of the migration; the job of network management is difficult enough without creating expectations impossible to fulfill.

This chapter defined Ethernet and Fast Ethernet as a networking technology, and included some subtle advice for designing effective migration strategies. This chapter discussed the physical, logical, and realistic performance limitations of the many Fast Ethernet variants so that you can design and implement Fast Ethernet. The next chapter discusses actual implementational problems and design issues, and shows sample designs for realistic migration scenarios.

CHAPTER 3

Fast Ethernet Migration Designs

This chapter defines the wiring and hardware required for installing or migrating to a Fast Ethernet network. Most of the chapter shows this within a framework called the premise, structured, or modular wiring standard. It is critical to conform to these standards, because components and test equipment for Fast Ethernet are designed and calibrated based on the assumption that a structured wiring infrastructure is in use.

The hardware requirements for standard Ethernet are minimal; they are not that much different for Fast Ethernet. This chapter describes the hardware commonly employed to construct Ethernet networks, and the physical channel limitations inherent in Ethernet. All Fast Ethernet variants, 100Base-T, 100Base-T4, 100Base-FX, 100BaseVG-AnyLAN, and switched designs are represented. Fast Ethernet represents a good migration strategy for greater bandwidth and sometimes reduced transmission latencies, because the hardware is simple and readily available. Furthermore, hardware is frequently interchangeable among vendors, platform types, and media variants. This provides compatibility between the equipment of different manufacturers, equipment of different vintage, and equipment of different Ethernet specifications—an important fact for the original construction of the Internet, which owes a lot to Ethernet technology, UNIX, and TCP/IP. Ethernet is often selected because it can bridge operating systems and ignore inherent differences in data and file structures. This basic simplicity promotes reliable interconnections. Relative to other network building blocks, the hardware for Ethernet is simple, reliable, and very powerful.

Success with any Fast Ethernet variant, particularly the 100-Mbps options, imposes a different level of detail and accuracy necessary for a successful migration. Although migration from a 10Base-T network installed two years earlier with a Category 5 infrastructure in whole or in part might be as simple as replacing just NICs and hubs, you do need to realize that moving to duplexed or switched architectures, or to 100BaseT or 100BaseVG, might also create transition problems. Specifically, the most frequent and easily resolved are the wiring infrastructure defects. These are typically minor, on the level of annoyance rather than a significant problem. In short, link-level cable testing pinpoints the wiring failures, which are usually found at panels, jumpers, and within connectors.

Testing calibration errors at installation or inferior installation can render the cabling unsuitable for Fast Ethernet. On occasion, entire sites do not conform to the Category 5 or even a Category 3 specification, despite the use of graded wire and connectors. However, you should know this before plugging in new NICs and hubs. A prerequisite for any

Fast Ethernet transition, even from (or perhaps especially from) 10Base-T to 100BaseVG, requires a systematic test of the wiring through to the patch panels to ensure that the wiring link can support whatever migration is proposed. That is why the wiring test is called a "link-level" test: because you want to check the link to there and back. Even though 100Base-T4 and 100BaseVG do not require the greater signal-to-noise ratios inherent in Category 5 wire and work with the lesser Category 3 and Category 4 wire, test for the availability of the necessary four pairs. 100Base-T4 and 100BaseVG aren't just about eight spare wires, but actually four pairs of two wires twisted together. It is a mistake to patch (or cross-link) different sets of wires to complete pairs, because this will defeat the noise-cancellation properties inherent in twisted-pair wire.

More to the point, the reason for 100-Mbps Ethernet is often lost in the false promise of more bandwidth. While many network performance bottlenecks stem from too much traffic on the wire, you might not increase overall usable network bandwidth by migrating to Fast Ethernet, because you merely free up one bottleneck and relocate the traffic buildup to other bottlenecks at servers, clients, bridges, routers, or in the extended infrastructure of a large enterprise network. You might refer to *LAN Performance Optimization* or *Enterprise Network Performance Optimization* for models and statistics that are useful in the analysis and design of better LANs and large networks. Although this chapter shows some of these same techniques, realize that the material is focused on a successful and strategic migration from standard Ethernet to Fast Ethernet.

TIA/EIA 568 TSB 67

The most interesting new standards for networking were derived from telephone wiring schemes. They are known as the Electronic Industry Association/Telecommunications Industry Association (EIA/TIA) recommendations 568 and 569 for premise wiring. It is important that you realize that these premise wiring definitions are recommendations, not standards, and not a point of law or legal requirement for building wiring. Because telephony, WAN service, and LANs have traditionally represented different personnel development tracks, most technical and managerial personnel do not have experience with both voice and data communication standards. It is time for consolidation, because the physical technologies are merging and the services provided by both are much the same.

The technical service bulletin (TSB) 67 fills in the gaps with a premise wiring recommendation simultaneously broad enough for telecommunication and data communication. The recommendations are useful for network wiring and the TSB provides measurement ranges for effective testing. Because Fast Ethernet is likely to include both UTP and optical fiber, the next sections discuss wiring requirements and techniques. The following illustration shows the important details for UTP with Fast Ethernet. This image may not be as illustrative of vertical and jumpered components in each lobe wire, or *horizontal channel connection* as the EIA/TIA terms it, but it does reinforce the importance and limitations of the recommendations. Figure 3.1 illustrates these specifications.

Copper

Twisted-pair is most commonly used for Ethernet in new installations. There are a number of types of twisted pairs commercially sold at different price points, marked for simplicity, and in common usage. All twisted-pair wire is fundamentally the same in that two wires form a pair and are twisted about each other. Each twisted-pair cable contains two, four, or 25 such pairs. Some twisted-pair sets may also bundle four pairs together with a drain or ground wire, and this is combined with other sets into a single cable for consolidating PBX, phone, and data connectivity. Specialty cables may contain other numbers of wires, or bundle optical fiber and armor into the cable. Some twisted-pair cable is shielded; this wire is the common recommendation for IBM Token-Ring and ATM on copper. The shielding is a wire mesh designed to reflect and contain high-frequency signals to minimize internal crosstalk and external interference. Types of twisted-pair vary by "grade" or "category," a measurement system based on the gauge and quality of the wire, the number of twists per foot, the tightness of the twists, and manufacturing savvy.

The EIA/TIA has ratified the premise wiring standard known as EIA/TIA-568. This provides precise guidelines defining communications wiring for data, image, and voice. EIA/TIA 569 provides guidelines for high-rise and commercial buildings. There are also related technical service bulletin (TSBs) related to actual field practice; the most dominant addenda are TSBs 39, 40, and 67. Forthcoming bulletins will reference data wiring within modular furniture, co-resident power and communication wiring, and how to build Ultra Fast Ethernet and ATM infrastructures.

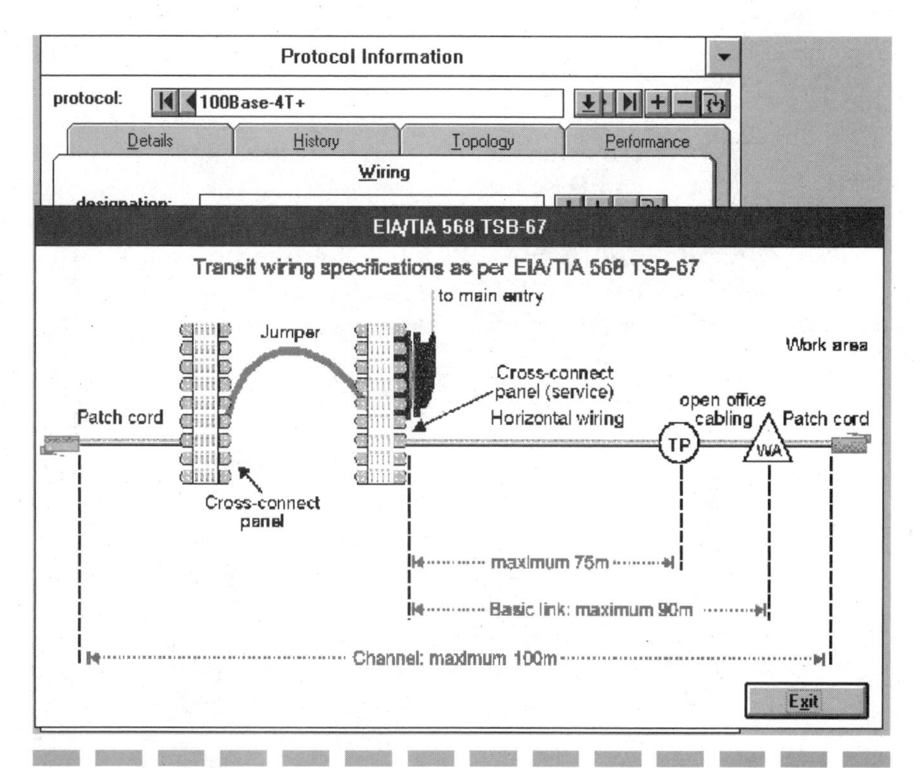

Figure 3.1.
TSB 67 premise wiring limitations. (*Network Performance Institute*)

The current premise wiring standards formalize Category 3 100-W unshielded twisted-pair (UTP) wire with four copper pairs as suitable for transmission speeds to 10 Mbps, Category 4 100-W unshielded twisted-pair wire (also with four copper pairs) as suitable for transmission speeds to 16 Mbps, and Category 5 100-W unshielded twisted-pair wire (also with four copper pairs) as suitable for transmission speeds to 100 Mbps. Realize that transmission speeds may be a fraction of the true signal speed, and that the slower signals on separate pairs are combined together to create a higher transmission speed. For example, 100Base-T transmits at 133 MHz, while 100Base-T4 and 100BaseVG transmit at 33 and 30 MHz on each wire pair. Although most network protocols only require two pairs, the standard is for four pairs to provide some future flexibility; quartet signaling may provide increased performance ranges impossible with just two pairs. Shielded twisted-pair (STP) with 150-W dual pairs is rated up to 155 Mbps, and is thus suitable for the low-end ATM cell relay over copper. STP-3 is comparable to Type 9, but is often used for Ethernet 10Base-T in older Synoptic LattisNet as conversions from the original IBM 4-Mbps Token-Ring.

Information regarding rack mounting and the nomenclature and requirements for rack-mount equipment and punchdown blocks is addressed only superficially by the EIA/TIA standards. Racks, panels, and cabinets are defined in a little more detail by EIA-RS-310-C (EIA-310-C), EIA-310-D, and the International Electrochemical Committee (IEC) as IEC 917. These standards present environmental performance requirements in terms of climate, shock, vibration, seismic activity, and electromagnetic interference (EMI). Additionally, neither the EIA/TIA 568 nor the 569 addresses the issues of collocation of data, voice, video, and power lines within integrated wiring solutions. "Integrated wiring" means raceway wiring components as manufactured by Leviton, Wiremold, AMP, and others.

Although the technology is reasonably well-established for collocation of power and other telephone wiring, integration and collocation with high-speed networking has neither been tested nor codified by any formal standards, or covered by any EIA/TIA revisions or technical service bulletins. My experience has been that it doesn't work well. It is chancy even with 100Base-T4 and 100BaseVG, and a nightmare with 100Base-T. What often happens is that the data wires in large bundles leak signal onto the power line, and as the large bundles of data lines are broken out and the lines get longer the signals tend to interfere with the furthest data lines. The signal gets weaker and jitter increases further from the 100Base-T hub, and hence gets more susceptible to interference. I have only seen it on outbound signals coresident with 120-volt lines; I suspect that ordinary 220-volt Asian, African, and European power lines would create more problems.

There is very limited knowledge about what works with power, telephone, and high-speed data collocation. Use prudence to maintain a separation of power lines and high-speed data wiring, as you do not want to create a serious infrastructure problem and performance disaster at design or integration. EIA/TIA 568 recommends a minimum of 0.5 m separation.

You need at least UTP Category 3 for 100Base-T4 and 100BaseVG-AnyLAN, although Category 4 and 5 will also work. 100Base-T (100Base-TX) requires two pairs of Category 5 or four-pair wiring that meets 100Base-T4. UTP itself is designated as either UTP-3, UTP-4, or UTP-5. This recognizes that the wire is unshielded, while the number usually refers to the EIA/TIA categorization. For example, UTP-5 is Category 5 unshielded twisted-pair. STP-3, which is a higher-grade shielded twisted-pair wire bundle, might or might not conform to a Category 3 or Category 5 specification; you will need to test each link. It is important to recognize that IBM STP Type 3 represents a different designation system from Category 3 wire. More often than not, vendors specifically designate the wire as "Category *X*" to make sure you know their products conform to EIA/TIA specifications.

UTP wire itself is typically 24-gauge copper wire with a PVC or Teflon jacket. Incidentally, a better grade of UTP-3 performs better than a low-quality version. It is true that UTP-4 is more expensive and performs better than UTP-3 if you extend the data rate on the wire, but the difference is not usually relevant for Fast Ethernet. Furthermore, shielded wire provides improved service in marginal installations, but should not support longer run lengths than unshielded wire bundles even though crosstalk is minimized, signal disruption is less, and signal attenuation is decreased. Again, the issue is that Fast Ethernet only requires a minimum threshold for performance and does not work better with a higher-grade wire infrastructure.

STP performs no better than UTP, even though it has better insulation against insertion capacitance and resistance to low-frequency electromagnetic noise—which means that it shields the signal and is rarely affected by external interference or internal crosstalk for the intended applications. The point is that the EIA/TIA specifications are intended to circumvent the market confusion and spurious claims for the many different types of wiring. For these reasons, the EIA/TIA specifications almost always refer to unshielded-twisted-pair as the common datagrade wiring medium for a premise wiring infrastructure, and network wiring specifications based on these standards should not be thwarted.

In fact, such nebulous issues of wire quality (above and beyond certification) are only important when a wiring infrastructure is not built with care. You do not need any margin of quality to run Fast Ethernet. As long as the infrastructure conforms to the corresponding EIA/TIA designation and is installed correctly, while punchdown blocks, modular connectors, wire itself, and path lengths conform within the appropriate Fast Ethernet limitations, the network will work. Forget about any thought that more quality in materials means a better network—it usually has no benefit within the current technology.

Rigorous manufacturing specifications and true adherence to EIA/TIA wiring and connector ratings will help ensure a reliable network and a generally accepted conformity among different lots of wire and different manufacturers. Shielded wire provides dramatically improved service and longer run lengths than unshielded wire bundles because crosstalk is minimized, signal disruption is less, and signal attenuation is decreased. The transmitting characteristics for an STP or UTP wiring bundle are comparable, but notably different. Although you can substitute STP for UTP, you are unlikely to generate any benefits in the near term from this more-expensive material unless the UTP was experiencing an unacceptable level of noise or crosstalk in a given cable run.

The signals on twisted-pair are dc current fluctuations. Twisted-pair Fast Ethernet is actually two pairs; one for transmission, and the other for reception. 100Base-T4 and 100BaseVG combine all four pairs for a single transmission or reception. As such, the advantage with 100Base-TX is the easier migration to a duplexed transmission scheme. These pairs cross over (usually internally to the hub electronics) so that transmitter circuitry operating on dc current connects to the receiver circuitry, and the receiver connects to a transmitter for each node pair to form the ring architecture. Although this may appear as a two-lane, two-way highway, the transmit pair is really only a continuation of the receive pair to complete the logical standard Ethernet bus structure. This is one of the vestiges of Ethernet that remain in Fast Ethernet. Transmission is unidirectional, unless you move to a duplexed transmission scheme. The stronger and more stable outgoing signal has a tendency to distort the weaker incoming signal, as Figure 3.2 indicates.

These parameters determine the ratings for twisted-pair. 10Base-T requires EIA/TIA Category 3 wire, and 100Base-T (100Base-TX) requires

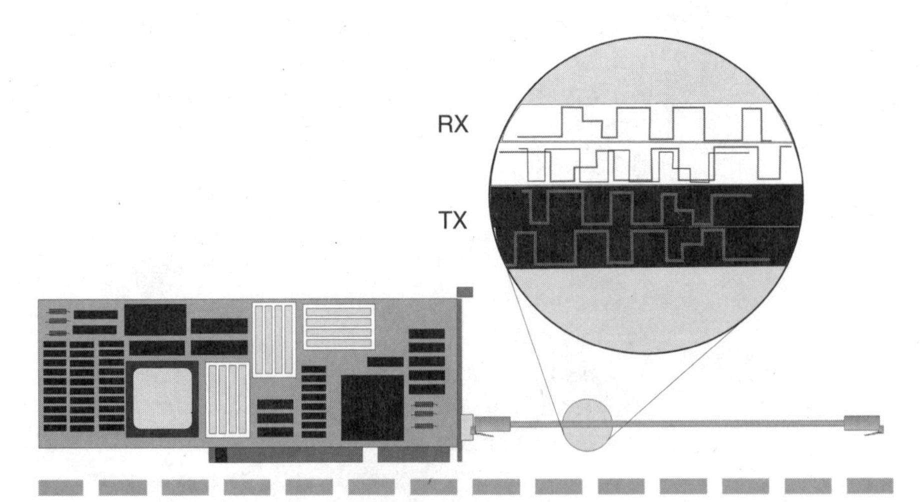

Figure 3.2.
Outgoing signals have a tendency to overwhelm and distort incoming signals with crosstalk.

Designation	Speed	Attenuation	Impedance	NEXT
Category 1	1 Mbps	4.3 dB	not measured	Not specified
Category 2	4 Mbps	8 dB	84-113 ohms	Not specified
Category 3	10 Mbps	11.5 dB	100 ohms	22.7 dB
Category 4	16 Mbps	7.5 dB	100 ohms	36.6 dB
Category 5	100 Mbps	24.0 dB	100 ohms	27.1 dB

Figure 3.3.
EIA/TIA wiring categories.

Category 5. Because 100BaseVG-AnyLAN and 100Base-T4 use quartet signaling and four pairs of wires, neither is so finicky about the quality and grade. Figure 3.3 shows the relevant performance differences between categories.

You can bridge preexisting 10Base2, 10Base5, or 10Base-T into a Fast Ethernet. Most 100Base and 100BaseVG wiring hubs include a 10Base connection, unless a 10Base-FL or 100Base-FX fiber connection is substituted for it. Remember, Ethernet specifies a bus design for coax and a star cluster design for twisted-pair. This becomes important when scoping the physical coverage of a Fast Ethernet network. Cable or fiber traverses a facility, and special bridging hardware interconnects stand-alone segments and twisted-pair hubs and extends the geographic reach of the network. Figure 3.4 shows a typical network configuration with a simplified representation of the Fast Ethernet hardware interconnected with a standard coaxial Ethernet. This is neither a physical nor logical representation of a network architecture, but rather a metaphorical cartoon. It is useful to abstract the concept of a data network to something commonly seen.

A Fast Ethernet network consists of a powered hub, lobe cables, and powered NICs in terminating computer equipment. Terminating computer equipment consists of devices, mainframe hosts nodes, servers, printer access ports, and user workstations. Older, coax-based networks are built with cable connectors, including both section end connectors and coupling "barrel" connectors (male-male), terminating resistors, coaxial taps, and transceiver drop cables. Powered components for a coaxial network are limited to transceivers (especially multiport units) and NICs at each node location. Bridges, gateways, repeaters, and other specialty equipment only expand this basic transmission capacity.

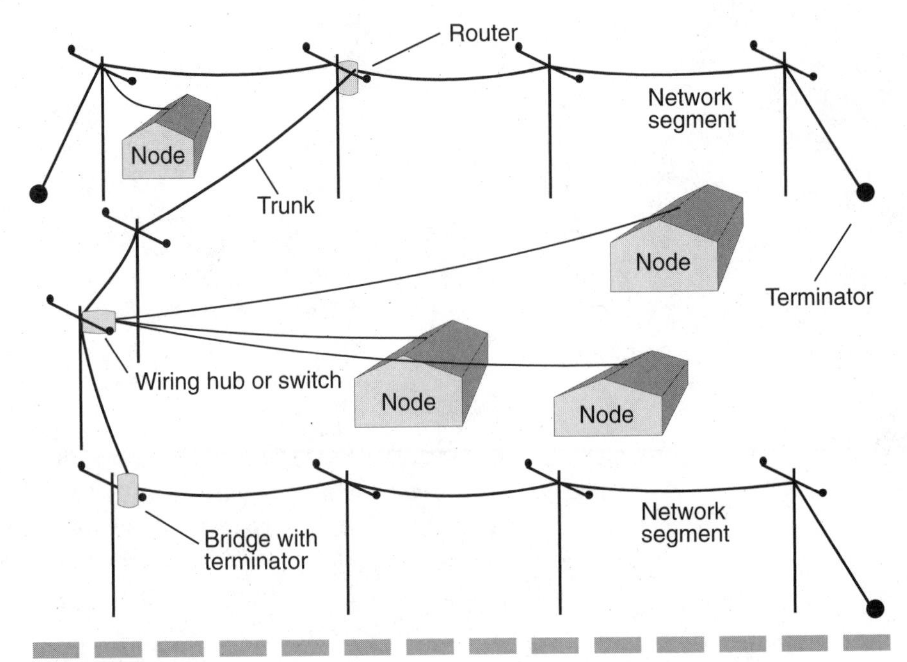

Figure 3.4.
Representation of an Ethernet network.

Conceptually, Fast Ethernet is very simple. The twisted-pair or optical fiber provides two-way simultaneous channel connectivity between communicating parties, which are the nodes. The channel is the data communications pathway. A packet or Ethernet frame is the transmission vehicle. While Fast Ethernet is a one-line country telephone network (only a single transmission can proceed at a time even in a switched architecture), the transmission protocols (which are rules of privacy or silence) accomplish an orderly transition for deciding who can transmit. Figure 3.5 pictures this analogy.

Each node has transmission and reception hardware (telephone handsets, if you will). This hardware controls access to the communications channel (coaxial cable, twisted-pair wire, wireless, or optical fiber) and monitors traffic. Additionally, each node has hardware that builds the messages to match the required Ethernet frame format. The transmission/reception hardware is called a transceiver, for *trans*mitter/re*ceiver*. This is analogous to the earphone and microphone. The traffic-control hardware is called an Ethernet controller, NIC, adapter, switch, or hub, and this is analogous to the box and dialing mechanisms of a phone system; it is built into each network adapter and each port in the central hub or switch. The

Fast Ethernet adapter connects to the workstation hardware via the computer bus and interfaces with system software. These two components form the interface between the computer and the network.

Often these two transceiver/controller hardware functions are built together as part of the workstation hardware rather than as separate pieces. The controller and transceiver, when bundled together into a single unit such as a personal computer adapter board, are called a network interface card (NIC) or a hub. In fact, the 100Base-T or 100BaseVG hub usurps the functionality of the controller and transceiver. This raises reliability and lowers hardware cost.

Backbones

Any Fast Ethernet technology represents a magnitude in performance improvement over an ARCNET, Ethernet, or Token-Ring LAN. Even when the fundamental network does not have bandwidth bottlenecks, the basic packet delivery time is faster. Hence a migration from "legacy" LAN protocols to Fast Ethernet represents an improvement. However, the savings from 40-ms transmission times to 2 ms is small in comparison to the 600-ms processing times on client workstations or network servers. Yes, Fast Ethernet is faster, but a transaction that required 1390 ms for a round trip on the old network will require 1314 ms on the new hardware. The difference may not matter at all. However, when LAN

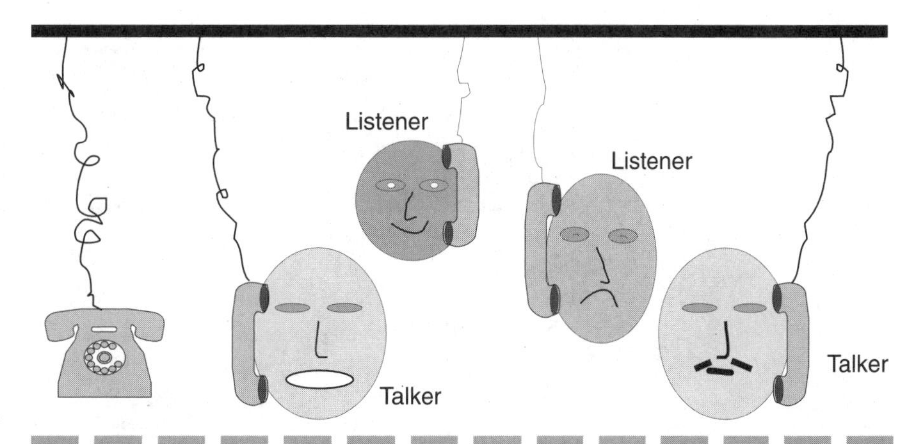

Figure 3.5.
The party line telephone analogy.

transmission delays are in the 300-ms range, saving 600-ms roundtrip on 1200 ms represents a 33-percent increase in speed. This is more than half a second of waiting time saved for a user. However, complex network structures will require routing or switching at the workgroup level.

Sample Designs

This section describes what is practical and necessary for implementing a Fast Ethernet network throughout a floor of a typical modern office building. Every network and environment is different. It would be great if we could categorize network designs like fingerprints. In any event, Figure 3.6 shows a typical floor in a modern high-rise building. You are going to see this blueprint again and again, because it is the basis for many case studies.

The building, like most high-rise buildings, has steel columns and reinforcement in the core columns and external corners. For the most part, the floor is quite open, and you can tell this is a modern building

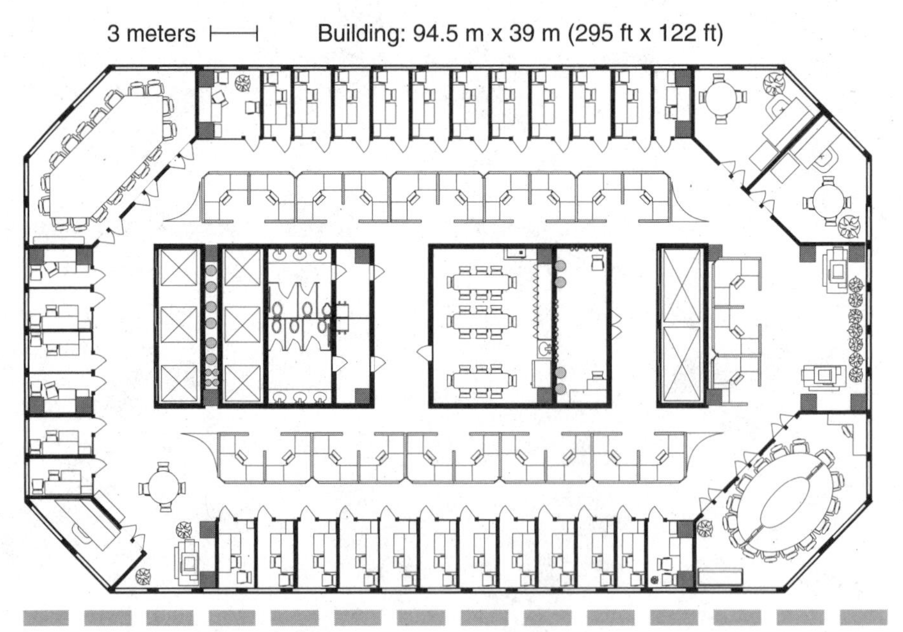

Figure 3.6.
A floor in a typical office building.

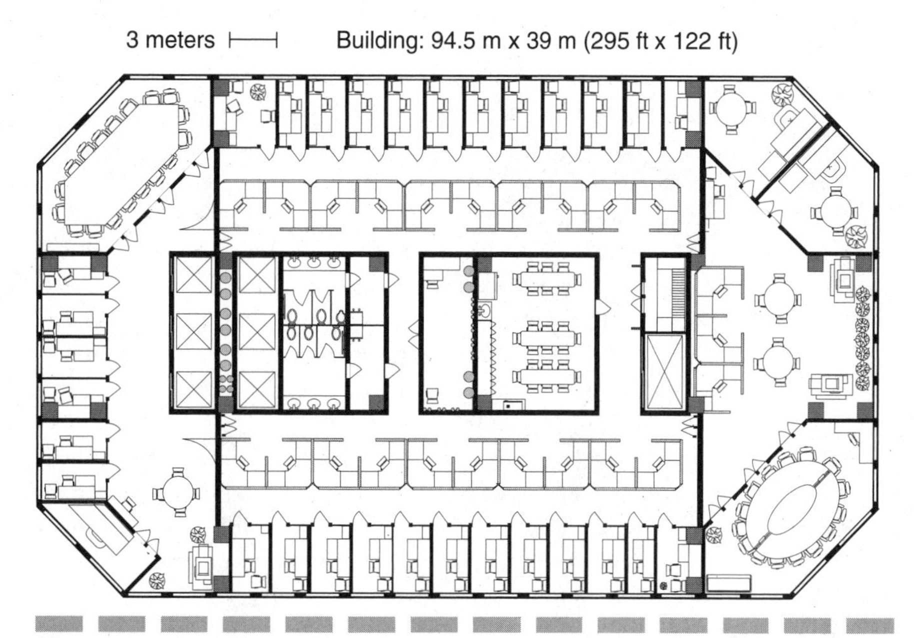

Figure 3.7.
A high-rise building with fire doors and fire zones.

because the wiring conduits and building services are in the core area adjacent to the lunch room rather than run through the elevator shafts. You might also notice the two sets of passenger elevators off to the side of the core, and larger service elevators near the wiring and service room. Note that the service room is not fully outfitted for the convenience of data communication and telecommunications wiring. In fact, electrical and HVAC are located here as well. Some buildings partition services and create separate areas for plumbing, electrical, fire control, sprinkler lines, voice communications, data communications, and HVAC.

Note the absence of fire doors in this figure. Most newer towers are partitioned into quadrants, and any firewalls are apt to be concrete block with self-closing steel fire doors. These structures are likely to be as impermeable to RF as they are to smoke and fire. This creates secure zones and may force you to create multiple zones as blueprinted by Figure 3.7.

Because the wiring closet is centralized in this building, the longest path from the wiring closet is no more than 60 m, or 66 m including the extra risers. Although the building is less than 100 m at its longest, you still need to run the communication wires in conduits that are firestopped; that adds to the path lengths. Larger buildings usually have secondary wiring closets to keep the lateral runs under 100 m, as was traditional with AT&T-compliant phone designs. The cubicles could be wired for infrared with two transponders, one for each side of the building. This might be easier than hard-wiring the cubicles with power, phone, and data. However, the cubicles will need to be wired for power and phone anyway. The issue of collocation of power with high-speed data communications has not been resolved yet. On the other hand, you are not going to get 100 Mbps on wireless just yet, either.

The presentation of sample Fast Ethernet wiring diagrams starts with the very simple, standard 10Base2 Ethernet bus structure on a typical floor of an office building. This technology was installed through 1989, when 10Base-T's UTP replaced coaxial cable as the primary Ethernet medium with powered hubs. The bus structure in the next illustration shows the backbone coaxial cables installed in the plenum above the false ceiling. Transceiver drop cables are not shown, but are typically pulled down through partitions and the vertical posts used to stabilize the freestanding partitions (see Figure 3.8).

None of this infrastructure, including the 25-pair transceiver cables, is useful for any variant of Fast Ethernet. The first step for migrating to Fast Ethernet is pulling twisted-pair from a central wiring closet to each office and cubicle. Although there is probably a similar wiring instructure

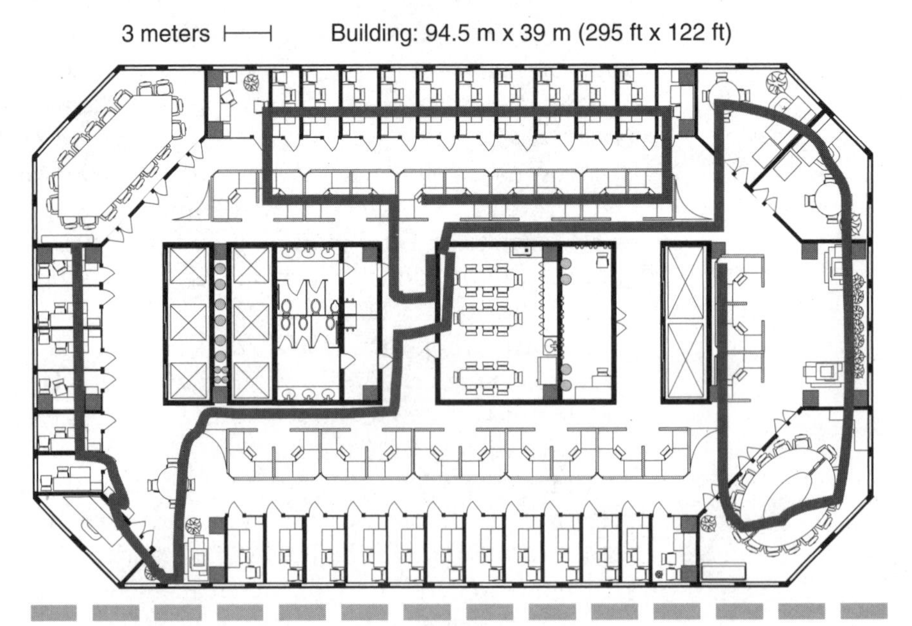

Figure 3.8.
Migration from a preexisting 10Base2 Ethernet LAN.

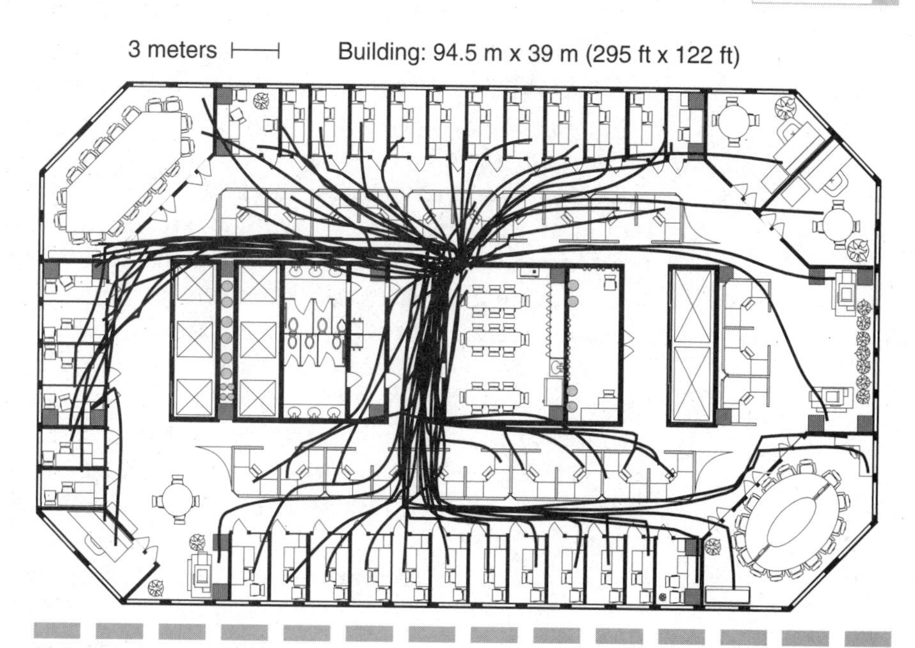

Figure 3.9.
The UTP installation required for Fast Ethernet.

in place for phones and intercoms, spare pairs may not exist for the data network. If you are planning a new telecommunications infrastructure, consider the need for four pairs for the phone system, at least four pairs for networking, and another four pairs for data for host connections, ISDN, or other services needed in the future, for each drop location. Many organizations wire five sets of four pairs for growth. Category 5 is the standard medium, although you can mix and match as needed. Figure 3.9 shows the distribution grid for the new wiring.

The cafeteria and wiring closet seem to coexist. Although any repeating or bridging hardware was probably stuffed into the ceiling or thrown into a closet not shown in the original 10Base2 blueprint, the space and security requirements for 10Base-T and 100Base necessitate partitioning the cafeteria and creating a locked wiring closet. Because phone service was probably wired through here, and the single service panel and small PBX wall unit did not seem so out of place in the cafeteria, consider now the importance of consolidating the wiring facilities. It may not be practical to reterminate the phone lines (because they already work), but you may want to collocate the new patch panels.

Although a preexisting 10Base-T network can be wired from a central location to each user and might look for all the world just like this, there are two limitations that typically prevent reuse of the preexisting wiring. First, 10Base-T required wiring that at best was Category 3, and it required only two pairs. Fast Ethernet requires two pairs of Category 5 or 4 pairs of Category 3. In the unlikely but very foresighted case that users were wired with 4 pairs of Category 3, you will only need to reconnectorize each location and create a different patch panel configuration for the four pairs. Figure 3.10 extracts the wiring layer from the prior diagram so that you can visualize just that wire grid.

New wiring typically costs betwen $200 and $300 for each line for pulling wire, installing panels and faceplates, and testing. Because this site requires 65 new lines, the basic installation budget for wire will range from $13,000 to $19,500. Some communication designers roughly estimate costs at $10 per square foot; installation could require $300,000 complete for PBX and network equipment. That translates into $4600 per user for PCs, servers, PBX services, wires, and phones. Transitions from 10Base2 to 100Base-T typically cost between $1000 and $1500 per port, because the basic computers are in place. Note that these preexisting computers may not have the horsepower to warrant the migration to 100Base Ethernet. Under these circumstances, consider a partial migration to 100Base Ethernet. Wiring just a single row of offices and extending a single wire to the server is a sensible intermediate step. Most users will not need the increased bandwidth, but rather only a few select processes and users will make use of it. If you want to bridge (or route)

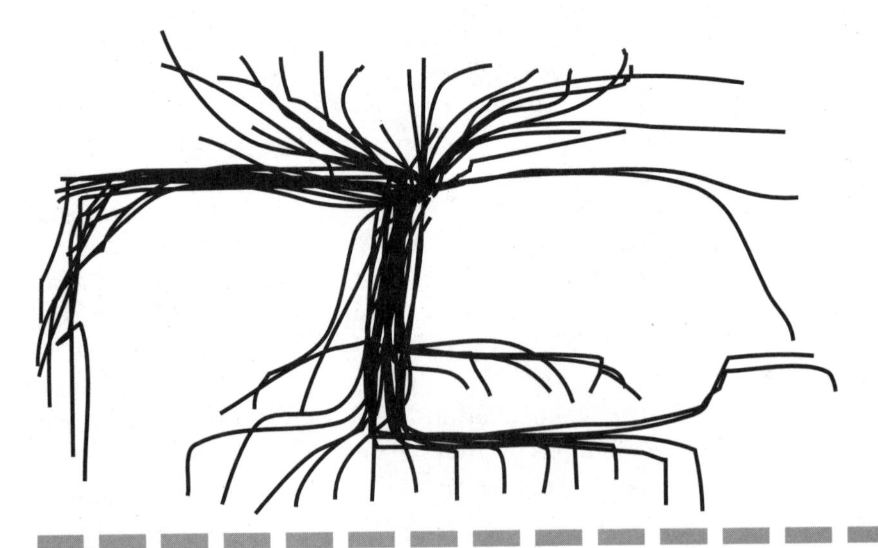

Figure 3.10.
The UTP wiring grid required for Fast Ethernet.

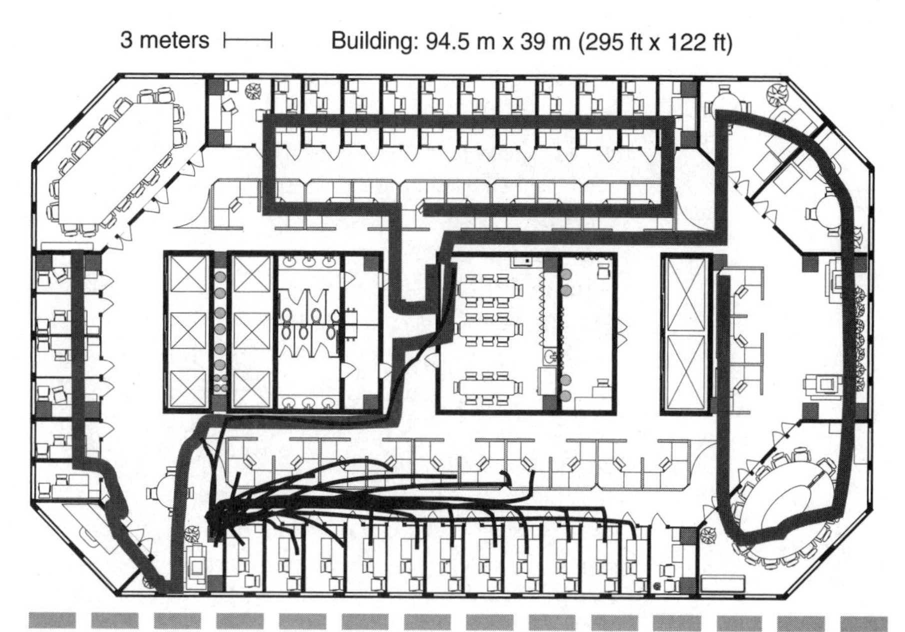

Figure 3.11.
The wiring grid required for a partial migration to Fast Ethernet.

the existing 10Base2 into the 100Base Ethernet, realize that the connections between the slower backbones and faster hub are a potential bottleneck, and any bridge or router is a long latency. Figure 3.11 shows a partial migration.

Single-floor sites are the exception rather than the norm. Most organizations exist over areas greater than 100×40 m and/or on multiple floors. Because Fast Ethernet limits connections to 100 m (200 m for 100BaseVG), the issue is usually not relevant for small organizations unless multiple wiring closets are required. Buildings with central courtyards often create some havoc because the wiring closet cannot be centralized. Larger organizations, such as a bank or insurance company in an office tower, would be limited to about 25 stories. Furthermore, a backbone of 100 Mbps will not be a lot of capacity. One solution easily overcomes these obstacles. First, risers (or links between wiring closets on the same floor) are installed as optical fiber. Second, all floors are not wired into the same backbone, but rather independently connected to a central facility, as shown in Figure 3.12.

Most of the larger vendors provide enterprise and workgroup Fast Ethernet hubs with either 100Base-F or FDDI ports. Each floor would be wired into a workgroup hub (or multiple ones) and all the backbone fiber would patch into the enterprise hub. Because the performance limitations for 100Base-F or FDDI are much the same, performance will be cheaper and faster without Ethernet-to-FDDI encapsulation or translation. This assumes that the backbone is an enterprise hub with a 3.2- to 20-GB backplane capacity. If you must design the network with a single backbone, FDDI is more robust in terms of latency, tolerance to load, and overall performance.

The 100-m or 200-m wiring limitation is certainly a design constraint when the Fast Enternet spans two or more buildings. You can build the two networks with various 10-Mbps repeaters and other hardware. This is the most frugal route. However, the linkage is limited to 10 Mbps and could cause a bottleneck. Ethernet or FDDI on fiber is the 100-Mbps solution for ranges to 2 km, and enhanced full-lasing FDDI connections can support distances to 30 km. In any event, you need a right-of-way to lay the cable. The logical configuration can be either a backbone or a series of connections from each site to an enterprise backbone. Figure 3.13 shows building-to-building connectivity by fiber.

These architectures show that you can intermix standard Ethernet with Fast Ethernet in the form of 100BaseVG, 100Base-T, 100BaseT4, and 100Base-F. The network does not need to be one thing or another, or

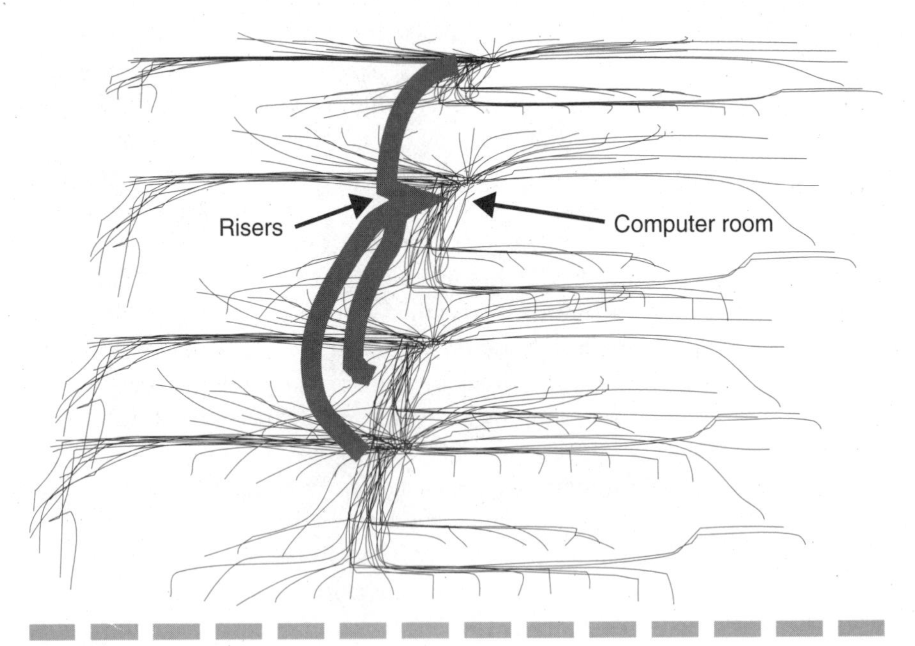

Figure 3.12.
Riser and horizontal backbone connectivity for Fast Ethernet.

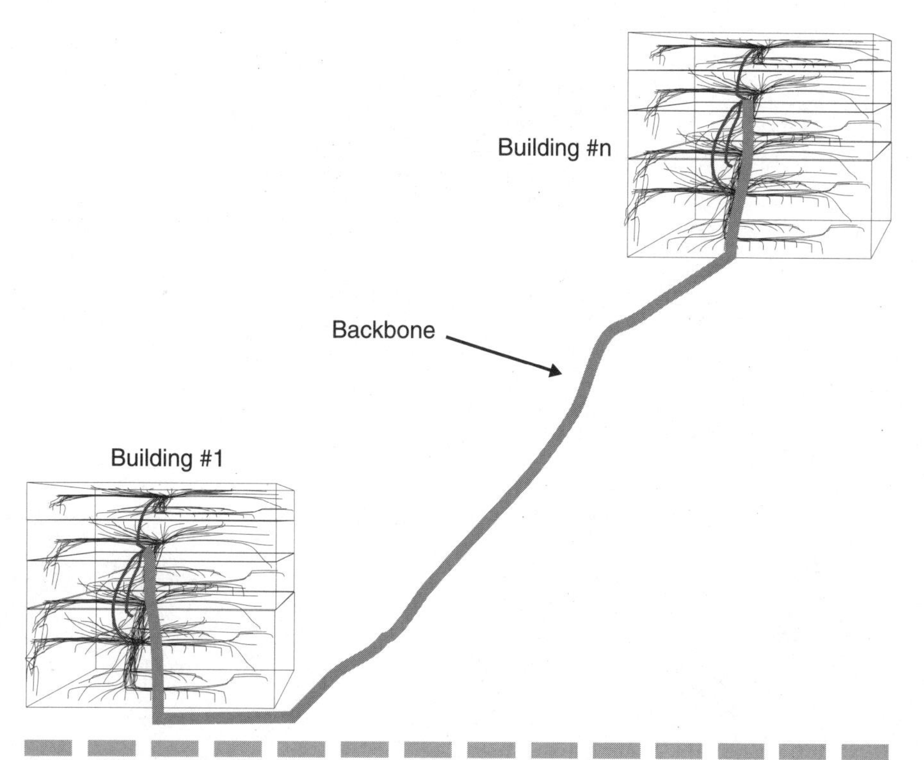

Figure 3.13.
Fast Ethernet on fiber for interbuilding linkages.

transformed in whole from 10Base2 or 10Base-T to Fast Ethernet; it can be a composite. Performance increases with Fast Ethernet are rarely a multiple of ten times over standard Ethernet unless you are migrating from very inferior situations. More likely the migration will provide from two to three times the usuable bandwidth, and make it clear that intermediate nodes and servers are new bottlenecks.

This chapter showed the fundamental mechanisms for Ethernet and the hardware required to actually build a network. Although new Ethernets are now almost exclusively built with a twisted-pair infrastructure, it is important to see that the Ethernet architecture owes much to its bus-based roots. In fact, all Ethernet LANS are limited by the timing parameters from the original specifications, with a network built of coaxial cable and transceivers. Although this may seem anachronistic, it ensures the continued viability and interoperability of existing Ethernet variants with newer technology. The next chapter details the hardware installation for Fast Ethernet.

CHAPTER 4

Physical Implementations

This chapter details migration, installation, and testing requirements for copper and fiber Fast Ethernet networks. I show how to pull wire or fiber, terminate it, test pairs, and deal with the myriad of physical issues that you will necessarily address when retrofitting 10Base-T to 100Base-T4 or 100BaseVG, building a new 100Base-T network in parallel to a preexisting LAN, or reaching other buildings and distant corners with 100Base-FX.

You can construct an extensive network with a complex, nonlinear tree-structured geometry to serve high concentrations of users, longer distances, and numbers of users beyond the network maximum of 1024 Fast Ethernet addressable nodes per subnet. Twisted-pair Fast Ethernet limits placement of nodes to within an absolute of 100 m (200m with 100BaseVG) from a wiring hub, as the last chapter detailed. Configuring a network as a single LAN may hold very restrictive limitations for some organizations. These limitations are circumvented by an intermediate node, such as the cascaded wiring hub, bridge, switch, gateway, and the router to interconnect two or more LANs into an internetwork. Also, fiber point-to-point links can, of course, create 2-km extensions, or 30-km ones if you interconnect sites with full-lasing FDDI.

Although some enterprise wiring hubs provide dense service beyond the maximum network concentration of 1024 nodes, you will likely need somehow to microsegment your network not only to stay within that node limit but also to provide adequate performance, even with duplexed, switched, or 100-Mbps Ethernet. The bridge and hub interconnect different subnets into one larger single "sub" network or network. The router, switch, and gateway interconnects separate subnets into one internetwork.

Physical Plant Limitations

Sharing computer resources and exchanging electronic mail through a built-in computer network is not reserved for well-funded start-up ventures in newly built industrial parks. Older buildings, unplanned sprawling complexes, and high-rise buildings can have networks, too. No building, no matter how old or how busy, need prohibit the installation of the type of equipment your organization wants. While it is possible that the layout of the building may constrain wiring efficiency, neither the building design nor lack of space will limit the kind of network that you can install. However, it might stretch the limits of Fast Ethernet and suggest the installation of 100BaseVG rather than 100Base-T, or the use of fiber for a backbone. In any event, as Chapter 2 showed, there is a strategy and a configuration suitable for any site.

If your building is of an older design, examine the premises more carefully; there might be additional costs for installing network wiring because the ceiling is inaccessible. Hanging ceilings or return-air plenums might be small or nonexistent. Existing pathways between floors or buildings might be unsuitable for new wire, or filled with old installations. In such cases, new conduits, cable raceways, or cable trays might provide part of the answer. Because polyvinyl chloride (PVC) cable produces a toxic gas when it burns, it is often banned from return-air plenums (which are the spaces above suspended ceilings). Fluorinated Ethylene Propylene (abbreviated as FEP, though better known as Teflon) insulated wire is more expensive than PVC, and exceeds most building codes without any restrictions. FEP cable sheathing is manufactured only by DuPont and one Japanese manufacturer. Because there has been a sudden surge in demand for Category 5 UTP, this demand has outstripped supply, causing shortages and price increases. Plan accordingly and get commitments for product well in advance of installation.

Twisted-pair, the cable most often existing in old buildings and now in new installations, might provide a ready means to migrate to Fast Ethernet. Qualify the wiring for its ability to adequately convey Fast Ethernet traffic on four pairs of Category 3-compliant wire or two pairs of Category 5. Assume growing demand on traffic loads even in low-use networks, such as those at a small business, a school office, or a satellite site. If growth, large numbers of far-linked machines in different directions, database access, or video imaging are required, it might seem uneconomical to reroute, add, or maintain twisted-pair. Rather, consider workgroup solutions connected by a backbone or enterprise hub. However, you will still want to create a premise-based, twisted-pair-based wiring plan.

Although twisted-pair must be run from central wiring hubs to each station, it does provide flexibility for use, management, configuration, and debugging. If the network is localized, when a contractor has to drill holes through walls, ceilings, and beams in order to route single cables through narrow plenums, it often seems disruptive to pull standard twisted-pair wiring bundles from office to office from a central wiring closet. However, the EIA/TIA concept of a modular and centralized wiring plant based on UTP is providing a plethora of parts and options and an ethic of building a uniform and flexible infrastructure.

Locations of computer rooms or wiring closets might be suboptimal but possible in many locations. Select the location with the best cable access and most stable climatic environment, and one that will provide security to meet design requirements. Computer facilities are best if locked and if the temperature remains constant. Overheating and low humidity, which increase static, cause component failures. Climate-control equipment generally solves these problems. However, these concerns are best addressed during the planning stages rather than as retrofitted necessities. The same conditions should be provided for secondary wiring closets, as they frequently contain wiring hubs and more complex networking computer equipment.

Consider how best to schedule installation. If cabling conduits and UTP are installed as part of new construction, the work is just one more item for the building contractors. Cabling an existing facility might require working around physical limitations and the office hours. If your organization cannot be closed during the network installation, the contractor might have to coexist with employees and customers. Additionally, scrap, dirt, and constant traffic might create a need for frequent cleanings. Acquire a good vacuum cleaner with extension hoses. A vacuum is a good tool for removing dust from circuit boards and grime from workstation air filters and inlet cooling ports on all electronic network components.

Remember one last point: Wire is often installed within a ceiling plenum that is about 3 m above the network connection for most nodes, as the prior chapter detailed. When planning wire runs from blueprints, add the necessary height drop or you will find that estimates are significantly short, and the network might be longer than planned. While 3 m (or 6 m, representing both the up and down heights) might not mean much in terms of buying more wire, accumulations of such extra lengths could push the lobes beyond specification.

Power Protection

Most network managers are aware of the danger of lightning striking outside wiring and sending a jolt down the line to damage anything unfortunate enough to be plugged into the outlet. Most managers are also aware of power surges and spikes that can cause premature failure of computer chips and magnetic media. Network wiring can also acquire induced or live voltages from proximity to power lines, not to mention the possibility that network wiring can be accidentally switched into live power. Some have experienced the slow degradation that occurs during a power brownout when voltage levels decay. Everyone knows the effects of a power blackout. The problem for the network manager is what can be done about these disabling situations. Wiring errors are particularly relevant, as data and voice lines are routed in pairs to offices. The higher telephone ringing voltages, although not fatal to people, are often damaging to LAN equipment. You also have to worry now about ISDN or other carrier lines connecting into LAN equipment. The higher transmission signaling speeds makes Fast Ethernet more susceptible to transient surges; the equipment is just as sensitive as any other networking equipment. Typical power problems for networks include:

- Power surges
- Power spikes
- Power sags
- Power brownouts
- Blackout
- Lightning strikes
- Stray voltage
- Static electricity
- Circuit overload
- Device startup surges

Also, radio-frequency interference (RFI) and electromagnetic interference (EMI) can travel through the power lines, with disastrous results to sensitive computer equipment. The damage from such events is not limited just to damaged computer equipment, data loss, and downtime. There are also lost opportunity costs, aggravation from frequent crashes, and additionally, the certainty that other damage will be spawned. A crash can destabilize a chemical reaction or a long clinical test, damage patient records and lose critical information, or ruin a part in a computer-controlled manufacturing process.

It is important to weigh the value of uninterrupted service and the cost of downtime. Frequent downtime affecting order entry, order processing, or accounts payable might be of little consequence, but it might represent the germinal stages of a bad customer-relations problem. The cost of the computer equipment is apt to significantly exceed the cost of protection, although the purchasing decision rests on the risk that the organization is willing to assume.

There are solutions to these problems. Specifically, filters screen out electrical noise from the incoming power, while suppressors level out the wave crests of power surges. Isolation equipment provides capacitance to level out the wave troughs of low power. Voltage regulators filter the electrical power so that the voltage is stable. Standby battery supplies and uninterruptible systems provide the same emergency power function. Uninterruptible power systems (UPS) provide electricity automatically when the sensing apparatus detects power problems. Because UPSs provide temporary emergency power only, most network managers allocate UPS capacity for an orderly and rapid server shutdown; monitoring software can track the performance of a UPS and initiate an orderly file-server shutdown when power reserves are depleted. If a UPS should support continuous operations, acquire a generator that will handle the circuit's capacity, including the starting surges of several machines. By the way, check the batteries in your UPS units; they are typically gelled-electrolyte units with a service life of two years.

Blueprinting electrical and network wiring and careful labeling of twisted-pair bundles can minimize errors. You want to keep power lines and high-speed networking lines at least one meter apart to minimize signal interference and NEXT (near-end crosstalk). This is still true for modular furniture with built-in racetrack conduits. One important note to realize is that most surge suppresser technology is based on a metallic oxide capacitor that degrades with each voltage spike. These should be replaced—at least tested—annually, before they fail to protect equipment. Most UPS systems include surge suppression; these suppression components might lose effectiveness and need replacement too. Commonly available electrical computer protection devices include:

- RF and EMI filters
- Surge suppressers
- Isolation transformers
- Voltage regulators
- Hot standby batteries
- Uninterruptible power supplies (UPSs)

Electrical Load Interaction

The initial current draw of a computer system or other electrical devices can create a disruptive load for critical and other network equipment. Photocopiers and laser printers, for example, can draw up to 3000 watts or 35 amps each, enough to draw down the available power so that a server, hub, or switch will crash. The best alternative is a UPS to level the power-clipping peaks and augment sags. While it is desirable to provide every node with a UPS, sometimes the organization cannot afford it, even though units for PCs have dropped below $100 (or $250 for workstations). Critical equipment should be protected, as shown in Figure 4.1.

When it is not possible to buffer network devices other than the critical ones—servers, replication nodes, intermediate nodes, and hubs—a less-expensive alternative (only when there is a large number of nodes) is to install an isolation transformer or power conditioner. These two different devices level the surges and buffer the sags up to about 120 ms. Problems that last longer than that, such as low voltages and blackouts, will not be avoided; you need a UPS for that. In small office environments, it

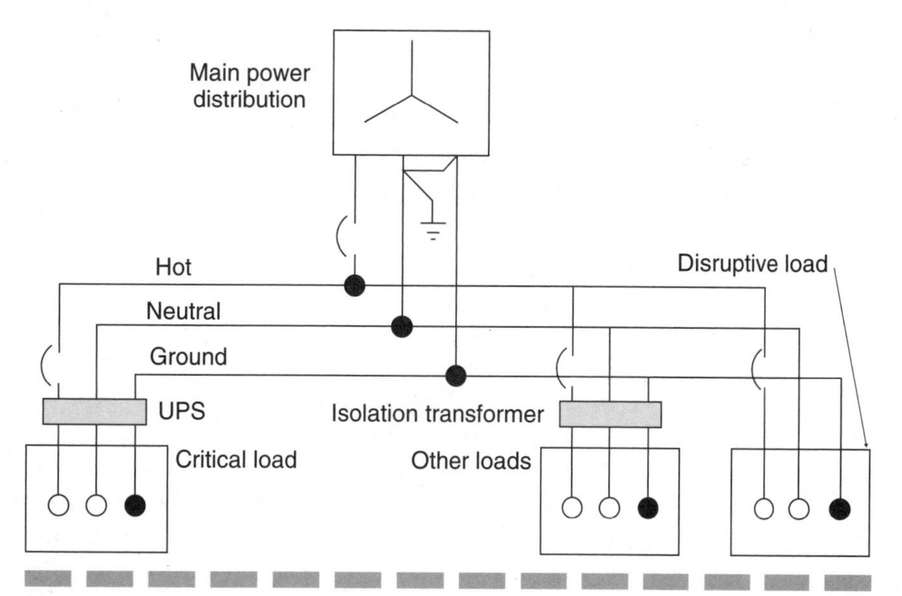

Figure 4.1.
Critical loads require UPSs, and other loads should be isolated with power conditioners or isolation transformers.

is often less costly to outfit each node with a UPS rather than wiring the main power circuits with transformers and either rewiring or installing special isolated lines.

Ethernet Compatibility

Ethernet provides some measure of compatibility. Not only does it offer a certain mechanical consistency, but many vendors sell products to interface disparate "Ethernet" components. Compatibility is often used as a selling point. However, this doesn't always mean that Ethernet-compatible equipment will transmit to other Ethernet-compatible equipment. This is particularly true with Fast Ethernet variants 100Base-T4 and 100BaseVG; although similar in wiring requirements, they are just not interoperable without a bridge between them. A bridge, or even a router, is a good solution for interoperability, but do not overlook the havoc added in terms in the latency, as described in Chapter 2.

This can lead to the potential for interfacing problems. While the physical media and even some of the transmission standards might be the same or similar, software discrepancies (at the NOS level) make functional incompatibilities. It is best to negotiate an overall networking plan to minimize these problems. If you can't, or arrive on the heels of a network reorganization and interconnection, use a bridging or gateway mechanism (sometimes called a linkage product) to connect totally separate networks for enterprise-wide communications. Consider the view of "Ethernet" presented in Figure 4.2.

Ethernet is defined as a medium and an electrical configuration as well as the media access method (e.g., collision detection). Variations might be "Ethernet-compatible," but not Ethernet-comparable without a special bridge. This means that these variants will communicate to a standard network through a specified mechanism; it does not mean that these alternative networks are electrically compatible. This is particularly relevant with regard to 100Base-T, 100Base-T4, and 100BaseVG-AnyLAN, all which require an attachment unit interface (AUI) for either 10Base2 or 10Base5 for traditional connectivity and compatibility. An ISO bridging or gateway mechanism, sometimes called a linkage product, connects totally separate networks for enterprise-wide communications.

In order for any third-party vendor products to function correctly with the existing network, all transmission and reception hardware must synchronize timing; NIC crystal clocks are not always accurate.

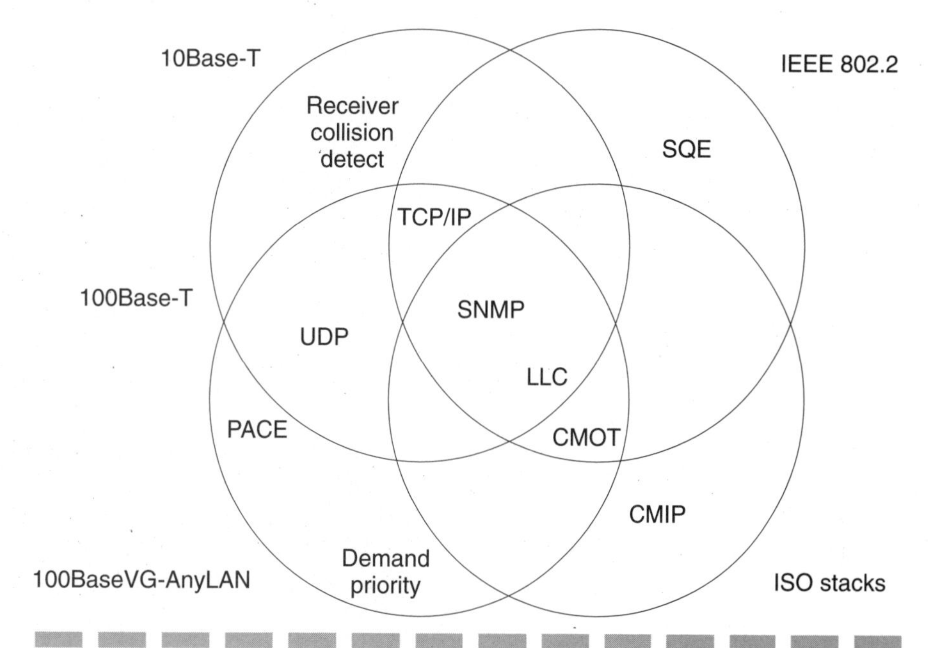

Figure 4.2.
Ethernet variant compatibility.

Subtle differences will become more apparent as the network grows in size and loading factors; these differences should be understood and the ramifications explored prior to actual purchase or installation.

Likewise, secondary sourcing of so-called commodity items might breed incompatibilities later when a network is upgraded. This is typical with many $29 Ethernet NICs. By the way, many of these low-cost ISA NICs are very good and comparable to $99 Eagle NICs. Less costly adapters might not support standard SNMP agencies or other features that may be important to network management. In switched environments, this network management tracking and analysis becomes very important. Servers and high-performance nodes that can create more than 2 Mbps of traffic should be upgraded to PCI-bus adapters so that the adapters can generate 10, 20, 30, or 40 Mbps of traffic. It is senseless to upgrade to Fast Ethernet if you do not increase the power of the transmitting nodes.

Also, some 10/100Base-T NICs do not support speed auto-sensing or 100Base-T and 100BaseT4 wiring options. The problem works for hubs and switches too. This might limit their immediate utility or future compatibility. Be aware that prime vendors might substantially alter their equipment or software over time for any number of reasons, including bug fixes, better performance, support for SNMP and RMON, a new sales strategy, or an attempt to undermine other vendors.

In any event, get verified customer references from both happy and unhappy customers. Assess compatibility and the eventual upgrade and customer-support policies. Vendors must support not only their own products, but also other network items that will interface with them.

Because Ethernet is a physical specification for a communication network, this network design only specifies the media and the electrical transmission standards; it does not define file transfer, mail systems, or other upper-layer protocols. These processes are supplied by NetBIOS, IPX, TCP/IP, and other such ISO layer 3 definitions; differences in the programmer's implementation of IPX cause one major incompatibility. Possible Ethernet discrepancies are:

- Physical channel dissimilarity
- Media dissimilarity
- Number of wires required
- Transmission preamble and delay incompatibilities
- Electrical signal differences
- Electrical timing differences
- Transmission speed differences
- Buffer size limitations
- LLC differences
- Heartbeat support incompatibilities
- Error handling discrepancies
- Acknowledgment differences

Different vendors supply their own interpretations of Ethernet and especially NetBIOS and NetBEUI. While Ethernet should be robust enough to provide access to all devices, in practice slight differences are indeed likely to create transmission problems. Such incompatibilities are more prone to happen on saturated networks. Different networks should isolate different types of equipment.

Not only are there many Ethernet transmission speed variations (for example, 1, 1.2, 2, 3, 10, 20, and 100 Mbps) and seven wiring versions (Thinnet, Thicknet, 10Base-T, 100Base-T, 100Base-T4, 100BaseVG-AnyLAN, and fiber), and an erratic migration to OSI stacks and MIB support, but each equipment and component vendor can redesign the functionality to minimize costs or maximize features. Small details, such as controller buffer parameters, NIC drivers, or how upper-layer software device drivers will access physical devices, can cause difficult-to-trace problems. Furthermore, vendor-supplied components can vary in small ways. While these minor deviations might not disrupt communications, incompatibilities might mean machines from different vendors can share a network without the ability to talk to each other. More ominously, such small differences can cause serious application-level shortfalls. For example, application software might not recognize the network printers, could crash suddenly and sporadically, could corrupt data files, or could fail to load into the PC or workstation memory.

One last point: When selecting NICs and intermediate nodes, make your life easy. Select equipment from as few vendors as possible and have exact spares. Should an NIC fail in a file server, you do not want to spend several hours reconfiguring the switches on the replacement, dealing with IRQ, DMA, or base memory-mapping issues, or have to rebuild and rebind the NOS. What you really want to do is open up the server, swap an NIC with an identical unit, close the case, and reboot. The same holds true for bridges, hubs, modular hub cards, switches, routers, and any other critical parts. Simply stated, in crisis mode all you have time for is a one-for-one equipment swap, and you can then reboot the device from the master boot floppy or swapped-out hard drive.

Planning for Installation or Migration

Installation requires a good plan. Migration is doubly complex because you need to maintain some level of functionality during the transition or else plan for complete switchover within a single day. I find that an orderly and slow transition creates less havoc and fewer weekend surprises that delay the switchover to another weekend. Network blueprints should be detailed enough to represent all the network hardware components required for a complete installation or migration. If the vendor (like an outsourced cabler or product vendor) does not supply this or provide complete network parts and installation services, it is vital to coordinate the arrival of all the network components. You will need a complete bill of materials. The network will not work without NICs, hubs, or any number of other components, not the least of which is functioning network software that works with the new hardware and at Fast Ethernet speeds.

The installation can certainly be staged, because many installation procedures can be completed concurrently. Just about any missing part, no matter what its simplicity or general insignificance, can become a critically needed component. Therefore, plan your installation timetable with an eye to the steps in network installation, and coordinate delivery of components to ensure a smooth process. Figure 4.3 illustrates the major critical steps and a logical progression for network installation. Each step encompasses several individual stages not included in the figure. Coax, for example, requires cutting to length; applying end connectors, couplings, and terminators; and installing the actual cable in its final location.

It is also relevant to create a project management time chart, sometimes called a Gantt chart, to show the relative sequence of events and the order in which they must occur. These charts also show the critical paths for network installation and what tasks can delay completion, as Figure 4.4 shows. A spreadsheet works, although complex products benefit from load-leveling and resource commitment tests that a real project-management tool provides. By the way, just about any task or step can hold up installation completion, but critical steps (usually beyond your control) include arrival of materials and installation of wiring. Incompatibilities between software and hardware and version issues represent another significant stumbling block, however one that is usually difficult to plot on a chart. It becomes a delay that you will want to build into your management-planning model.

Installation is a mechanically simple process, but a challenging experience if the technology is new to the organization. I suggest that Category 5

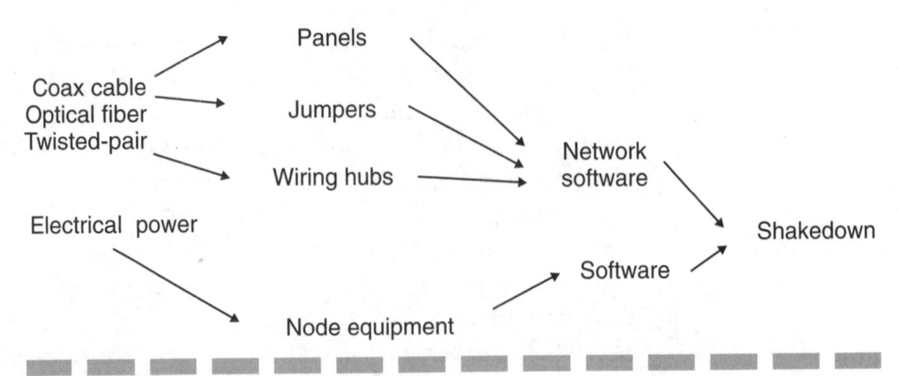

Figure 4.3.
Critical paths for network installation.

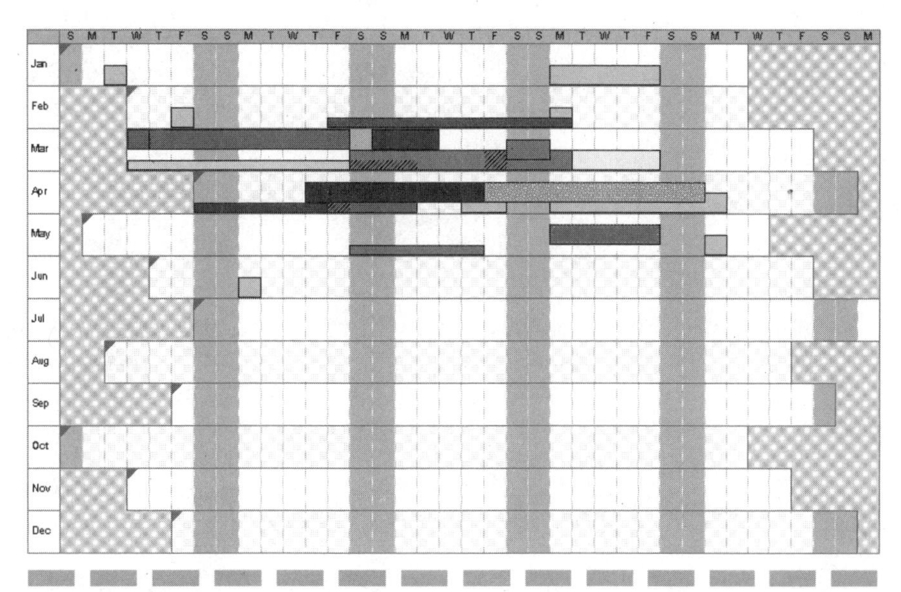

Figure 4.4.
A project management time chart.

is very different technology from 10Base-T or Token-Ring STP. Do not be lulled in a false sense of competence by the physical similarities. Sufficient skills might not have been developed to adequately identify and repair network problems on Fast Ethernet. Realize that you might not have the tools, experience with the tools, or access to tools for Fast Ethernet. In some cases, existing network-management software is wholly incompatible with Fast Ethernet equipment. Switching hubs are difficult to test with a protocol analyzer, even one that supports Fast Ethernet, because problems might be split across multiple segments that you cannot view all at the same time.

Therefore, consider the benefits in hiring the vendor or an experienced contractor to string coax and tap this cabling. Gather as much outside experience and expertise as possible to augment the internal network administrative team. The installation procedure requires a shakedown period. Premature acceptance of a new installation is unwise, and making or disseminating promises of network availability before most of the problems have been resolved will devalue the perception of the network administration team. Budget sufficient time for overruns, problems, and missing components. Be certain to negotiate preconditions for any final acceptance with the vendors. An installation might provide marginally acceptable operations, but fail to perform the level of service desired. To forestall any future problems, define clearly what your expectations are. This is critical for successful vendor relations.

Preparing for the Physical Installation

Installation of a new network goes smoothly if you gather the preliminary blueprints, have all the equipment and components before beginning, and if you proceed in the order listed in Figure 4.5.

Installation Tool Kit

Fast Ethernet requires the same tool kit as the one needed for 10Base-T. Note, however, the tools are very different from 10Base2 or

Steps	Equipment
1. Create preliminary blueprints	CAD or network design software
2. Project plan	Project planner or spreadsheet
3. Order components	—
4. Schedule power/wire installation	—
5. String wire and label	Cable, wire, tool kit, labels
6. Secure wire	Cable ties, cable tie tool
7. Install connectors and panels	Tool kit, connectors, panels
8. Test infrastructure	Test equipment
9. Install nodes	NICs, transceivers, hubs
10. Load software	Software, working nodes
11. Test NOS functionality	SNMP, protocol analyzer
12. Test performance	SNMP, analyzer, monitors
13. Shakedown cruise	Run standard applications
14. Update blueprints	Preliminary blueprints

Figure 4.5.
Steps and equipment required for installation.

10Base5, because the wiring infrastructure is more akin to telephone wiring than cable TV. Installation requires a sizeable tool kit. The contents are enumerated here solely to provoke some ideas. This kit should contain:

- screwdrivers
- pliers (including needle-nosed)
- diagonal cutters
- a cable cutter
- a wire stripper
- a bright flashlight, preferably rechargeable
- rosin-core silver solder
- a soldering iron
- desoldering equipment
- crimping tools
- a shorting plug (a connector with a wire soldered to short the core conductor)
- various spare parts like probe tips, junction boxes, and wall plates.

If you are installing twisted-pair, you will need a punchdown tool (it is also called an impact tool). Some new equipment automatically installs a 25-pair wire to a punchdown block in one motion. Include specialty tools such as cable corers for vendor-specific or antiquated taps and connectors, and modular wire-crimping tools in the size matching your modular connectors. Note that RJ-11, RJ-22, and RJ-45 crimping tools are different, and even for the same connector there are variations that are convenient for UTP or STP installations.

Less obvious tools include ladders, a stud finder, walkie-talkies, and a radio. In a clever application of a standard carpenter's tool, the electric stud finder also locates active cables and wires. This can be useful to find ac lines so as not to route network cable too close. A portable AM/FM radio serves not only to relax hard-working installers, but also provides cheap testing of any audible EMI sources that might disrupt network communications. Given the location of most cabling, a ladder or scaffold is suggested; a special multiple-jointed ladder that acts like a scaffold fills both needs.

Note that falls from ladders are an occupational risk for network staff. Unfortunately, a fall from a ladder is not considered preventable until you or one of your staff falls and breaks an ankle, shatters a pelvis, or lands in the hospital with a concussion. Take this risk very seriously. A fall could cause grave personal injury, damage to network installations, and subsequent downtime. There is the additional loss of productivity while the person heals. There are also financial consequences like sick time salary, workman's compensation, and perhaps Occupational Safety and Health Administration (OSHA) inspections and fines. Acquire safe ladder skills.

The art of climbing ladders is not taught to most people nowadays; it is a skill from yesteryear. Furthermore, network management is often a white-collar profession and there is a blinding perception that computer-related jobs are safe. Learn to climb ladders properly. Park the ladder on level ground, lock the ladder's folding supports, and never reach out so that your center of gravity is not over the ladder. Never overextend your reach. Avoid the top step of folding, A-shaped stepladders. The T-bar in the ceiling you think will help you balance or support your weight will probably give way, because it is only attached to the cement or steel beams with a little wire. If you cannot reach, step down. Move the ladder. Remount the ladder and continue your work. Install tool caddies or hooks to keep important tools within reach, or provide belt-mounted or vest tool carriers. Avoid painter's extension-type ladders, because they do not provide stability. If you have to use an extension ladder, the vertical reach to the horizontal slope should remain at a ratio of 4 to 1. In other words, if you are climbing up 10 feet, set the base of the ladder 2.5 feet from the wall or support. Add a ladder stabilizer gripper to the top of your ladder for additional stability, as Figure 4.6 shows.

Consider the condition and utility of your ladders. Maintain ladders. Tighten bolts. Replace rickety ladders. Because network management often begins as an ad hoc process, the old maintenance ladder is often the only available ladder. Do not chance it; purchase a new ladder. Wooden rather than aluminum stepladders are inherently safer, because they do not conduct electricity. Fiberglass is stronger and is another good choice. Multijoint ladders that can assume a variety of articulated shapes tend to be more stable, versatile, and safe from falls, although they are usually made from aluminum.

Ladders are not the only tools to extend your reach. Library-style rolling stairs are a practical acquisition and inherently safer than ladders due to their larger silhouette and side railings. If the ceilings are unusually high, acquire scaffolding. Hydraulic lifts, sometimes called cherry pickers, often have safety cages. For industrial complexes, factory manufacturing floors, and space buildings with extended ceilings, obtain a rolling platform (perhaps even motorized) to reach those heights. Plat-

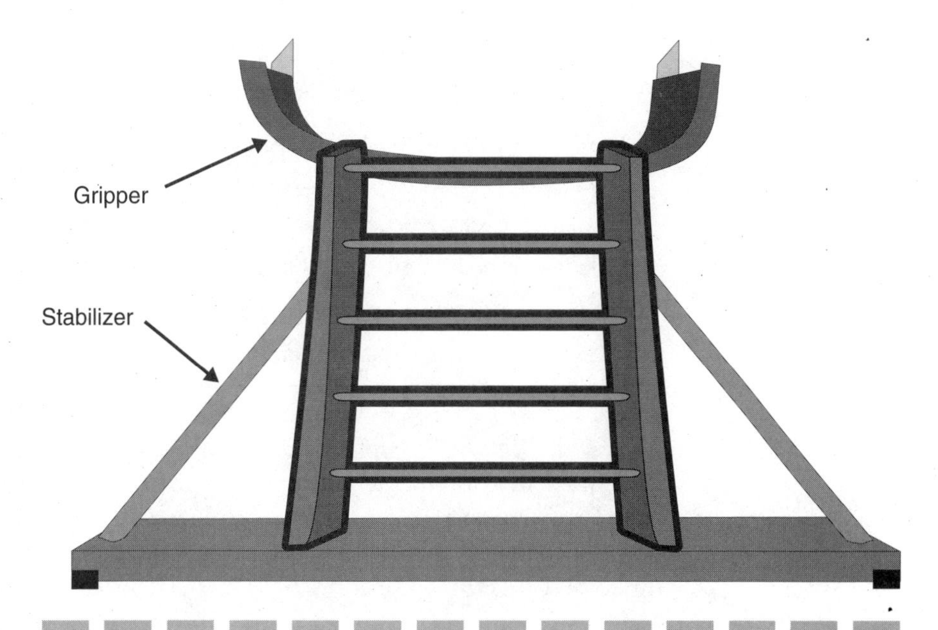

Figure 4.6.
A ladder safety extension attaches to standard ladders.

forms are inherently safer than long ladders and provide immediate tool access. Some people attach safety lines with quick-release clips to roof trestles as an added precaution. Remember that falling tools can injure bystanders below. Remember, too, that extension cords to power your tools are a risk atop the ladder both for the electrical potential and their chance of snaring you with a dangling cord. The next list summarizes practical ladder precautions:

- Set up the ladder on level ground.
- Lock the ladder's folding supports.
- Maintain center of gravity over ladder silhouette.
- Never overextend your reach.
- Do not move suddenly.
- Move deliberately and carefully.
- Do not balance on ceiling components.
- Move the ladder directly beneath your work area.
- Avoid the top step of folding A-shaped ladders.
- Select a ladder tall enough to safely reach your work area.

- Carry tools on a belt or vest.
- Avoid extension-type ladders.
- Add a stabilizer gripper to the top of the ladder.
- Maintain ladders in safe condition.
- Replace rickety ladders.
- Select wooden or fiberglass rather than aluminum ladders.
- Acquire scaffolding or a rolling platform.
- Attach safety lines with quick release clips.
- Ground the ladder.

Installations using optical fiber will require specialized tools that are not normally obtainable at hardware stores, and that are also quite different from those required for twisted-pair 100Base-T. This includes a fusion splicer, a stabilized light source with the appropriate light frequency (generally 850 nm), a light meter sensitive to the same frequencies, and a scanner or optical time domain reflectometer (OTDR). It is also good to have access to a protocol analyzer with sufficient capacity to decode Fast Ethernet at speed and full channel capacity, support all protocols and encapsulation technologies, and also be able to create traffic at the full 100-Mb/s rate to test the network and attached devices. A microscope is useful to check fiber endpoints for proper polishing. A polishing kit is inexpensive. An epoxy connector kit is necessary only if you manufacture your own connections or need to repair a damaged connection; consider obtaining one for all building sites.

Another tool specific to optical fiber is a signal detector that can indicate which fibers are active without cutting or breaking the fiber. A similar and inexpensive device is available for electrical lines; it lights up when in proximity to active electrical circuits. Obtain one sensitive to UTP and STP voltage levels (10 volts dc), telephone voltage levels (48 volts dc), and for ac power lines (110 volts ac and up). This later range is useful to correctly identify dangerous power lines before touching them.

A floor tile lifter for extracting raised floor panels might be another necessary choice. Walkie-talkies, cellular phones, or similar mobile communication devices provide a simple way to keep the installation or repair team in touch. If the site is large, two-way radios with a 3- to 5-km range are less expensive than cellular phones. If you are going wireless, consider adding wireless phones to the PBX before installing the LAN. A multitester and various other test equipment are also suggested.

If you plan to install any fiber, the tools are generally more costly, but not necessarily so. For example, Wavetek sells a handheld tool for approximately $5000 to test UTP integrity and conformance to the Category 5 recommendations. Although such a tool is expensive, the costs for downtime and getting a vendor in to effect repairs typically pays for the tool within 9 months for installations of 80 or more nodes. Wavetek sells an add-on kit for fiber that is only $900 more. This combined tool kit is much cheaper than buying separate copper and fiber test tools; you will save about $8000 by buying this single tool rather than two separate ones. Overall, the modular connector tools, cutters, crimpers, and powered punchdown tools will cost about $300. Brand-new cutting, polishing, and splicing tools for fiber run about $1000. The differential for adding fiber is only about $2000. You can cut corners by buying used equipment from jobbers and bring that difference down to a few hundred dollars. Fiber optic equipment has been around a long time, and big users like national carriers retire old equipment when it no longer generates a tax deduction and is superseded by faster tools. Figure 4.7 illustrates a robust tool kit.

Another highly useful tool is the electronic tape measure. This device sends out an infrared signal (because it is the same wavelength as wireless infrared) to measure distances within an accuracy of a couple of inches. It is particularly useful as a planning tool to estimate wiring requirements very accurately. When measuring cable runs, do

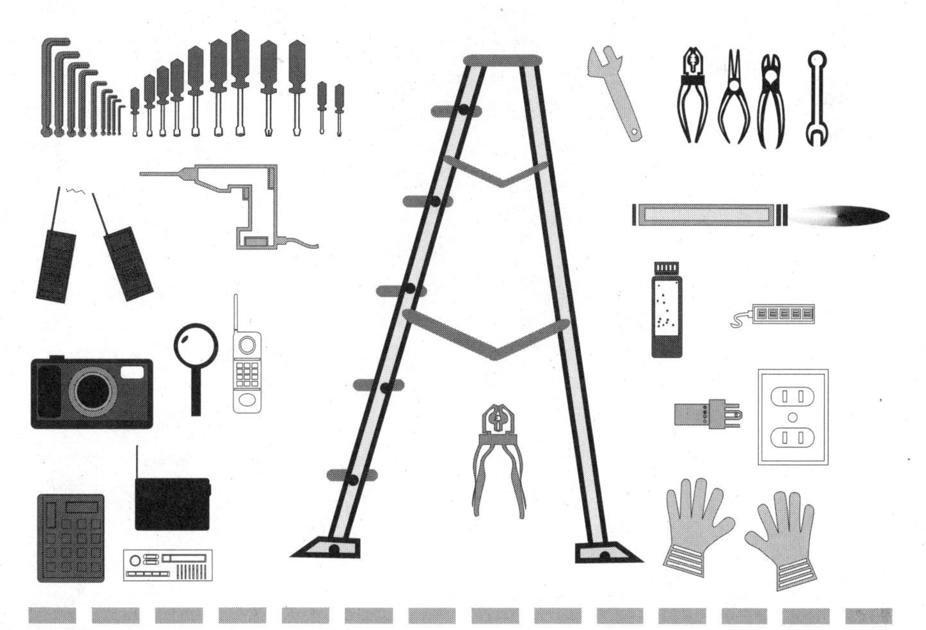

Figure 4.7.
Typical installation tools.

not overlook the vertical heights between nodes, wiring hubs, and other devices, and the channels where the wire or cables will actually be installed.

Cabling

Fast Ethernet is fundamentally plug-and-play. It usually works as promised and delivers the bandwidth enhancement sought when the cable infrastructure is installed correctly. When Fast Ethernet creates problems, it is either a design flaw or from cable problems. The cable problems are the more frequent because the 125-MHz Fast Ethernet signaling presses the limits of most cabling components. Fast Ethernet is predominantly a copper UTP medium, although fiber is an alternate choice. I would suggest that fiber designs apply FDDI instead of Fast Ethernet, because the performance is better and the cost is actually less at the current time. You also get a better backbone with more flexible connectivity and design options. Dual homing with FDDI becomes important as an automatic fail-safe in mission-critical environments. Some of the many cabling options are shown in Figure 4.8.

Of course, transitional installations tend to work better with a single protocol, and in that case, stick with the transitional mix of Ethernet and Fast Ethernet. You do not need the headache of tunneling or translating Ethernet packets to FDDI frames and back again. Use four-pair copper as a simpler transition, but use the TX standard if you need to install new cable. Because 10Base-T also used only two pairs (even though the specifications show four pairs wired in the connectors, of which two really get used), many installers cut corners and only wire two pairs. The other pairs might get used for telephones or split for secondary data lines. As such, the current installation might be inadequate for 100Base-T4 or 100BaseVG, which require all four pairs. While it is sometimes possible to rewire the patch panels and switch plates to enable all four pairs (there might very well be two or more spare pairs in the walls and ceilings for every drop), rewire and test a few connections for conformity before making the final plan. Some 10Base-T networks, although they work, do not conform to Category 3 and will not pass a scanner test.

When you have to rewire, consider what might be a sensible and future-looking plan. In most corporate installations I suggest Category 5, but if growth and bandwidth is of prime concern (and considering the minor differential in costs) go with fiber instead. Fiber is actually easier

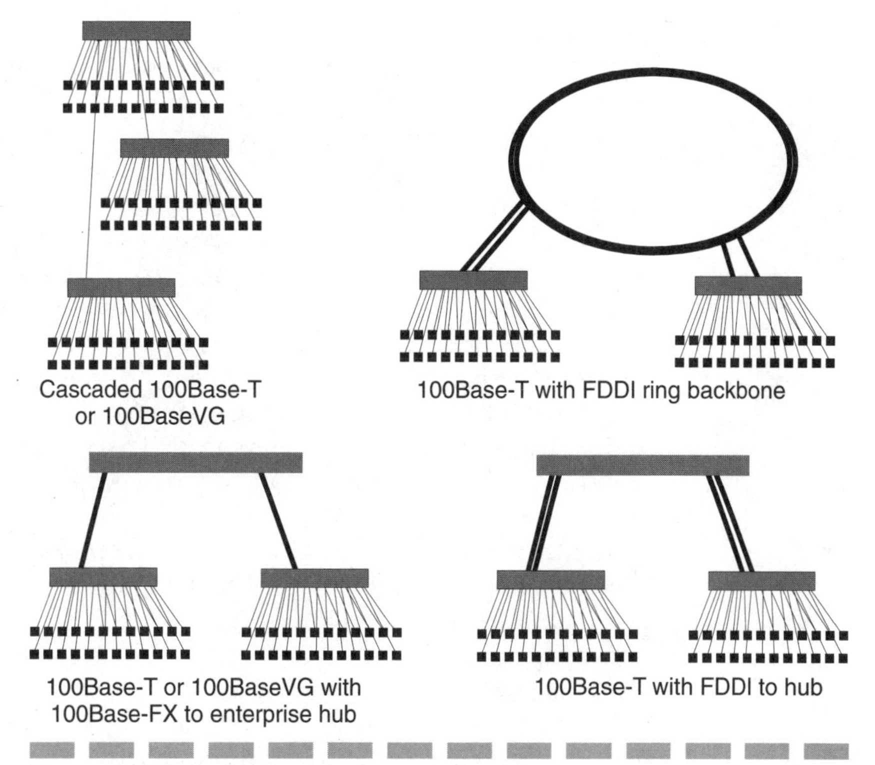

Figure 4.8.
Network cable options.

to build, maintain, and manage from my experience with FDDI. However, the focus of most Fast Ethernet products is on copper, and you give up some of the management tools built into most top-of-line enterprise devices. Most choices with hubs, switches, and extraordinary equipment are designed and sold for UTP. Many of the higher-end products include a fiber connection between hubs and switches, but it is the virtual LANs, switching, duplex transmission, and firewalls that are really built into the copper-based equipment.

High-Speed Copper

Twisted-pair cabling, connectors, and patch panels are available in many grades and categories, only some of which are suitable for Fast Ethernet. Ascertain the minimum requirements for your network and plan accordingly. Recycled connectors are not worth what you think you might save, and components not conforming to the EIA/TIA standards result in a fundamentally flawed installation that is high in crosstalk, signal interference, and impedance losses. If you did not get the message before this, most Fast Ethernet migration problems result from substandard installations.

The plenum-rated material jacketing most Category 5 wiring is in short supply and will be for several years, according to manufacturers and industry analysts. Although the shortage initially resulted from a fire at a DuPont Teflon resin manufacturing facility, this has been exacerbated by an extreme stockpiling demand for in-plenum Category 5 material. Only DuPont and a single Japanese supplier make the Teflon resins that meet fire code requirements. The alternate, generic Japanese products are not qualified in all counties and localities as meeting fire code rules. Some legislatures specify "Teflon" as the material required rather than the generic fluorinated ethylene propylene in-plenum cable. Check the rules first, and do not get caught flatfooted fighting a senseless legal battle with a building department; even if you win, it will cost dearly. Jumpers usually do not need to be fluorinated ethylene propylene, but again, check the regulations.

While excessive demands in conjunction with the real shortage have resulted in higher Teflon material prices, getting sufficient material really is not a problem if you plan your installation with enough lead time. Large jobs might require 2 to 3 months lead time for planning. On the other hand, you might find that the cost differential between copper and fiber (which is only in the wire materials) hub and NIC interfaces makes fiber backbones and fiber to the desktop a reasonable consideration.

If you are outsourcing the wiring installation, confirm the credentials and experience of the installers, and if at all possible, obtain an actionable guarantee for the quality and durability of the installation. Several large national installers provide paperwork that guarantees a functional Category 5 installation for as long as 15 years. I have seen these installers follow through replacing and repairing shortfalls as organizations have migrated from Token-Ring or 10Base-T to 100Base-T and discovered significant flaws one and two years after initial installation. Most failures have been in connectors and patch panels, or the actual connection of only two pairs instead of all four pairs, although some wire pulls needed replacement. Because the sales representative is not the same person who typically oversees the actual installation job, and you might build a good rapport with the sales representative that will not carry over with the job supervisor or cable technicians, it is important you know the details and installation process; you are the one who will have to follow up on details with the technicians who actually install the cable.

Transmission signal frequency is the determining factor on what cable you will need. Category 3 is sufficient for 100Base-T4 and 100BaseVG. Category 4 cable is an odd commodity, and is not really necessary unless your environment includes or will include 16-Mbps Token-Ring; use at least Category 4 or IBM Type 1 or Type 2 STP wiring. If you are installing 100Base-T or foresee a need for it, or are contemplating 155-Mbps ATM over copper, install Category 5 wiring. The alternative with Category 3 (or Category 4) is 100Base-T4, which is very similar to 100BaseVG-AnyLAN in that it requires all four pairs of wires for each node connection. You will also need NICs and hubs supporting the 100Base-T4 variant. In general, two-pair 100Base-T on Category 5 is simpler.

Selection of the wiring is not the only determinant to a functioning high-speed infrastructure. All connectors also must conform to the Category 5 standard as well. If you select inexpensive connectors and blocks, you are likely to see initial problems with impedance or possible corrosion at a later date. Substandard connectors, blocks, cross-connect patch panels, and jumpers create near-end crosstalk problems, impedance, and mapping problems. Additionally, the EIA/TIA 568 standards specify proper installation techniques so that the infrastructure will comply with the concept of premise wiring, but also function as a high-speed network. These EIA/TIA 568 standards (additionally the EIA/TIA 569 and assorted addenda) are the *lingua franca* for premise wiring. While important, they are not the last word. A lack of definition for rack mounting, power and data wiring requirements in modular furniture, and requirements for higher speeds on copper are still unresolved. Furthermore, the endurance of a Category 5 infrastructure has not been tested over long periods, which is perhaps a very good financial reason to pay the premium for a national installer with a 15-year warranty to wire your sites.

Cable Installation

Before installing any new or repaired wire, or adding twisted-pair, inspect it for visual defects. The wiring bundle should have no obvious physical defects like cuts, tears, or bulges in the outer jacket. Electrically, the individual wires in the bundle should conduct electricity without discontinuities. Each wire in a pair should be electrically isolated and not shorted out to the other. It is also very important to test that pairs are not separated. In other words, you do not want to mix a wire from a red striped pair with a wire from a brown striped pair; this defeats the noise cancellation of the twists and could spell crosstalk and excessive NEXT. Techniques for testing with a cable tester are presented later in this chapter. To install the twisted-pair bundle you will need the tool kit, the preliminary blueprint, the cable, cable ties, and colored electrical tap for labeling.

Category 5 wiring by itself does not guarantee 100-Mbps performance all by itself. Connectors, panels, and jumpers must conform also to Category 5 standards. Installation per recommendations is also essential. Proper installation means more than passing a continuity test. Installing unshielded twisted-pair cabling is a highly skilled art, requiring installation practices specified in industry standards. In some states, the Union of the Brotherhood of Electrical Workers (UBEW) or other trade groups have lobbied successfully for restraints on the legal right to install Category 5 wiring. In any case, not all installers are competent, and you should realize that Category 5 wiring requires significantly more installation care than voice-grade lines.

The primary standard applicable to cabling installers is the EIA/TIA-569. This is the telecommunication wiring standard for commercial buildings. A draft revision labeled SP 2840A is due to be issued soon as TIA 568A. ANSI, USOC, BICSI, ECMA, and other trade groups have also ratified this standard, and there are similar international standards issued by the International Organization for Standardization and the International Electrotechnical Commission as ISO 11801 and DIS-11801. These are basically the same as the EIA/TIA-568 standard for Category 5 issues; they do not cover all the material of the EIA/TIA standard. Most major suppliers of telecommunications components provide installer training based on these standards. The cabling issues that must be addressed by the installer include:

- Minimum bend radius
- Maximum pull force
- Jacket removal
- Untwisting of pairs
- Maximum cable length
- Avoiding sources of electromagnetic interference
- Cable termination methods
- Patch cord assembly

These issues are addressed in this and subsequent paragraphs. The TIA's draft SP 2840A, also referred to as Annex E to EIA/TIA 568, specifies that the minimum bend radius of Category 3, 4, and 5 UTP cable shall not exceed four times the diameter of the jacketed cable. This requirement is less severe than that published in the earlier EIA/TIA-568 specifications. Avoid exceeding the minimum bend radius when cable is pulled around corners or coiled for storage behind wall plates. Bend radius is important, because excessive bending of a UTP cable might change the geometry of the twisted pairs within the jacketed cable and cause crimps that alter signal transmission propagation speeds and accuracies. This can adversely affect the near-end crosstalk and attenuation performance of the cable.

Pull force is exerted on a cable when it is being installed in a ceiling, raceway, cable tray, or conduit. The effects of exceeding the recommended pull force include stretching the cable, which can change the geometry of the twisted pairs within it. This, in turn, can adversely affect the transmission performance of the medium. Although no standards requirement for pull force currently exists, SP 2840A recommends that four-pair UTP cables be able to withstand a pull force of 110 newtons (25 foot-pounds) without damage. The same maximum pull force should suffice for Category 3 and Category 5. You might need separate risers and intermediate pull boxes to stay within these limits. Pull force can be reduced with a lubricant. Water is usually sufficient, but tension reduction with silicone-based lubricant could be significant in multiple-bend pulls. Some installers have taken to talcum powder. This would include pulls where "bending" is caused by undulations in conduits and plenums or by tight corners.

Fast Ethernet lobes are very sensitive to irregularities in the physical media caused by kinking, stretching, binding, and general rough handling. Wire pairs in a typical Category 5 cable are composed of two insulated conductors very tightly twisted around each other. This twisting minimizes not only susceptibility to outside electrical noise, but also crosstalk from one pair to an adjacent pair. To meet Category 5 parameters, it is imperative that installation techniques be used that will maintain the integrity of this twist throughout the entire system. Problems that can be localized to the cable installation itself can usually be attributed to the disruption of the integrity of the twist. Stretching, a common issue with softer nonplenum cables, can actually cause the wire thickness to fluctuate from normal to thinner-than-normal, thus lessening the available signal path and increasing attenuation.

Kinking of cable is sometimes difficult to avoid. However, impedance mismatching and excessive near-end crosstalk (NEXT) are two problems that can arise from a kinked cable. Tight radius bends cause similar problems. For example, an impedance mismatch occurs when the cable is so tightly bent that the relationship between the conductors is disturbed, resulting in fluctuating capacitive and inductive characteristics at that particular point. SP-2840 requires that bends in Category 5 cable installations be no less than 3 cm (1.25 inch) in diameter. If binding occurs when the cable is pulled tightly around a sharp object such as a support beam, hanging ceiling hardware, or ventilation equipment, damage can range from a slight flattening of the cable pairs to complete sheath destruction and removal of individual conductor insulation. Problems caused can range from mild increases in crosstalk to open and shorted conductors.

When terminating UTP cable, you should only remove as much cable jacket as needed to terminate properly to the connecting hardware. This is the TIA's recommendation in SP 2840A, which allows for installer discretion, because various types of connecting hardware require more or less jacket removal. Minimizing jacket removal maintains a barrier that protects individual conductors from physical damage and electrical interference.

Category 5 cable pairs shall not be untwisted more than 13 mm (0.5 in) and Category 4 cable pairs shall not be untwisted more than 25 mm (1 in) at the point of termination, according to SP 2840A. There is no requirement stated for Category 3 cables. Common sense dictates that no more than 25 mm (1 in) should be untwisted. This is illustrated in Figure 4.9.

Complying with these cabling requirements maintains proper link performance to reduce near-end crosstalk. You do not want or need to strip the wire. Common punchdown methods (and tools) and modular connectors pierce the wire insulation as part of the installation process.

Maximum cable length is specified in the original EIA/TIA 568 standard as 90 m (295 ft) for horizontal cable runs, regardless of cabling medium. The maximum length for backbone cable varies by medium and is specified in SP2840A. This recommendation states that backbone cable length might vary by application. If the spectral bandwidth of the application that will operate over the backbone is greater than 4 MHz (IBM 4-Mbps Token-Ring is 8 MHz), the total backbone cable length should be limited to less than 90 m (295 ft) for UTP and STP A cabling. 100BaseVG supports about double that, of course. Jumpers or cross-connects at the wiring closet can cumulatively extend to 6 m, while wall connections to workstations typically cannot exceed 3 m.

Transformers and other energy sources can alter the impedance of the wire. Minimize the wiring bundle lengths, and do not leave tails of wire

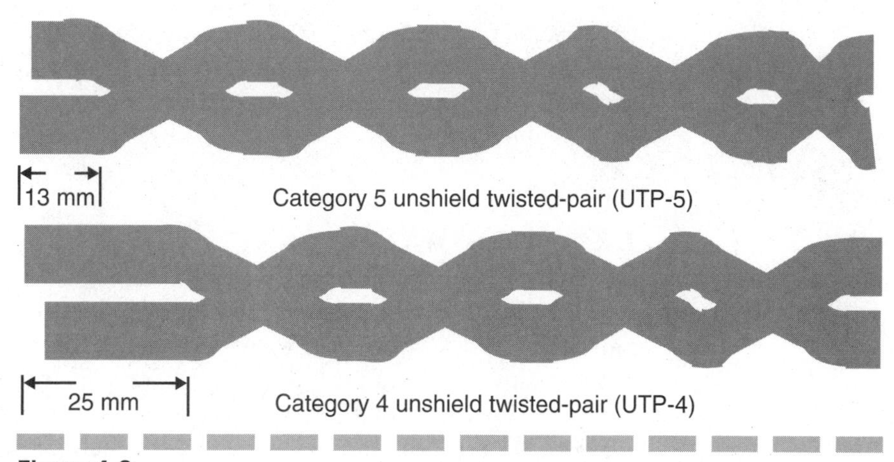

Figure 4.9.
Do not exceed the recommended limits for untwisted UTP.

(coils of extra wire) above the wiring closet; cut the wiring bundle to the shortest possible length, but leave sufficient slack for temperature-induced expansion and contraction, as well as for the possibility that wiring punchdown panels or office partitions might get moved a few feet.

If you anticipate that the distance limitations or external inference might cause transmission problems, select the highest quality of UTP. Do not overlook vertical cross-connects and drops when estimating cable or wiring measurements, because the extra lengths increase the potential that external interference will disrupt data communications. You can offset the extra length by substituting a higher grade of cable. For example, if the normal grade is Category 3, substitute Category 4. Shielded twisted-pair is available (at a substantial premium, as it is not as widely used and not as common). Category 5 and STP Category 5 will perform better for all network applications because crosstalk, signal degradation, and insertion to capacitance are improved; however, this extra service is not required for any current networking technology, and your network will not be better for it.

Until recently, the cabling itself was thought to be the primary cause of NEXT interference. However, current manufacturing techniques are producing very consistent, tightly twisted cabling that is capable of handling extremely high frequencies with minimal NEXT interference. Now the modular connectors and termination facilities are the points in the link at which the cable's structural integrity is compromised. The problems occur when the cable must be untwisted and spread apart to allow connection to a plug, jack, or punchdown block when interfacing to active or passive interconnecting hardware (e.g., patch panels, hub, concentrators, or switches). The more connecting and termination points that exist, the more susceptible to interference and signal degradation your system will be.

Labeling Bundles

As you string the wire, label it with some identifying marks. Twisted-pair wire is still color-coded—each pair with blue, orange, green, and brown with white—and each of the jackets for these pairs has a different color and hash mark to identify each individual set at each end of the bundle. Multiple bundles are confusing and should be labeled separately. Color-coded electrical tape at 15-foot intervals helps sort out the spaghetti. Medical supply houses also carry alphabetic labels useful to indicate more specific information. Fluorescent marking tapes provide exceptionally effective identification above false ceilings when you are perched on a ladder nine or more feet high; visibility is often impaired by ductwork, supports, and wiring.

Securing the Wiring Bundles

Once the wire is properly routed and installed in the cable trays, conduits, ladders, and J-hooks, or draped over the suspended ceiling supports, the cable should be tied with plastic ties to prevent damage.

Patch Panels

Most other forms of Fast Ethernet are modular and centralized through the hub or switch. The star configuration encourages installation of a patch panel for centralized wiring. Additionally, because twisted-pair networks and telephone circuits often share the same wiring, the common telephone wiring schemes are used. Punchdown patch panels separate and differentiate the 25 pairs of wires from each bundle. The punchdown panel is for cable termination. A hydra or separate jumpers connect into another panel for patching into the hubs or switches. This separate panel is not a requirement, but make the premise wiring extremely flexible; you can change connections by moving the modular jumpers around.

The patch panel is marked with the locations of the other ends of each pair (which is highly recommended), and each individual wire is snapped into a retaining pin. When actually wiring the individual pairs, maintain polarity and do not cross the pairs over; the wires should at all times be straight through from plug to plug. For network applications it is best not to overuse the wiring bundle. The following process is recommended for terminating Category 5 UTP cable on a 66 block:

- Mount the block's bracket on a properly prepared plywood wall surface or a cross-connect frame. This bracket is useful for cable management and slack storage.
- Route the cables through the bracket and distribute them evenly to the left and right sides—for example, six four-pair cables per side.
- Snap the connecting block onto the bracket.
- Remove only as much cable jacket as necessary to terminate the cable conductors. Remember to leave some cable slack behind the block and bracket for new terminations later.
- Dress the cable pairs through the fanning strip slots as pairs. Do not split pairs or mix them up. Both conductors of each pair should be inserted through the same slot in the fanning strip.
- Once the conductors are inside the block's fanning strip, split the two apart and terminate them onto the quick clips. Do not untwist the cable pairs more than 13 mm (0.5 in) for Category 5 cables and no more than 25 (1.0 in) for Category 4 cables.
- When dressing cable pairs through the fanning strips, alternate slots so that the first pair is pulled through the first slot, then skip the second slot, pulling the second pair through the third slot, and so on. Alternate slots when terminating to the connecting block. Cross-connect wires can be pulled to the empty slots; this covers commercial building path slots later. The alternate, empty slots will be used for cross-connect wire when necessary. Visually inspect the cable pairs before terminating.
- Terminate the conductors using an impact tool.
- Labeling might be done on the connecting block fanning strips, add-on designation strips, or cover.

This procedure will ensure that the pair untwist of any twisted pair of conductors will be less than or equal to the 13-mm (0.5-in) requirement of SP 2840A and EIA/TIA TSB addendum 40a. The recommendation of limiting the amount of cross-connect jumper wire or patch cords to 6 m (20 ft) is another important consideration to ensure proper cross-connect frame design and layout. Recall that the patch cords at workstations cannot exceed 3 m (10 ft). The recommended procedure for terminating Category 5 on a 110 connecting block is similar to that for the 66 block, except:

- Lace the cable conductors into the retention slots on the 110 wiring base. Do not untwist the cable pairs more than 13 mm (0.5 in) for Category 5 cables and more than 25 mm (1.0 in) for Category 4 cables. Visually inspect cable pairs before terminating wires.
- Terminate the conductors using a five-pair impact tool. This action will seat up to 10 conductors at one time, and will cut off the excess wire.
- Place a 110 connecting block on the head of the five-pair automatic impact tool. Align the connecting block over the appropriate 110 wiring base location and terminate the connecting block onto the base. The top of the connecting block is used for termination of the cross-connect jumper wire.

Wire technology, although it seems very industrial and low-tech, actually has evolved considerably in the last few years. Not only have modular components lowered labor content and increased reliability, the technology has also evolved to protect you from yourself. Color-coding and labeling is simplified with automatic labeling tools, laser-printed adhesive tags, and even special devices like wire clips for punchdown blocks. These clips, shown in Figure 4.10, are useful for two-pair Fast Ethernet configurations if you use just one block for both lateral wire and jumpers. You can also do the same thing with 4-pair Fast Ethernet if you alternate rows. I have seen this done, but I consider it too complicated and difficult to maintain. I use these clips to color-code areas, office wiring, or types of wiring; however, they are just one thing to do that doesn't get done if there is no secondary purpose to them.

There are two significant differences between 110 patch panels and 110 connecting blocks. A 100-pair 110 connecting block requires the use of cross-connect jumper wire to connect different cable distribution fields, and 110 patch panels require the use of stranded wire patch cord assemblies to connect the separate cable distribution fields from panel to panel. The procedure for terminating four-pair Category 5 UTP cable on a 110 patch panel is similar to that for the 110 connecting block, with the following exceptions:

Figure 4.10.
Slot protection is useful in dynamic operations, messy operations, or new sites to minimize stray wire shorts and lost connections. (*Siemons Wire*)

- Mount a rear wire manager behind the patch panel, preferably near its top. If mounted too low, it might obstruct access and make cable termination from the rear of the panel difficult.
- Route cables from the back of the patch panel through its coupler openings and loosely attach them to the wire manager with cable ties. Leave enough cable slack so that a coupler might be removed from the panel at a future date for reterminations.
- Lace the conductors into the retention slots on the 110 connecting block. Do not untwist the cable pairs more than 13 mm (0.5 in) for Category 5 cables and more than 25 mm (1.0 in) for Category 4. Usually inspect the cable pairs before terminating wires.
- Terminate the conductors using an impact tool. This action will seat the conductors one at a time and simultaneously cut off the excess wire.

- Snap the coupler into the patch panel.
- Once all of the cables have been terminated and couplers have been snapped into the panel, dress the cable slack behind the panel and tighten the cable ties.
- Labeling might be done on the patch panel face plate paper tabs and clear plastic covers that are supplied for each coupler.

Figure 4.11 illustrates a common punchdown panel. Jumpers are constructed to connect each pair on the panel into the wiring concentrator. When planning a 10Base-T or 100Base-T network, be certain to confirm that ac power for the wiring concentrators is available near these hubs.

There is an art and pattern to terminating the wiring at the wiring closet. It should be neat, but it is also useful to conform to standard installation practices. Specifically, Figure 4.12 shows how to connect pairs to the horizontal wiring blocks.

It is important to label all the connections carefully and maintain up-to-date records for each wiring pair. Risers (cables that go to other floors) get lost very quickly. Label them as well. If you interconnect a hub or network node with a telephone circuit, the network device will shut

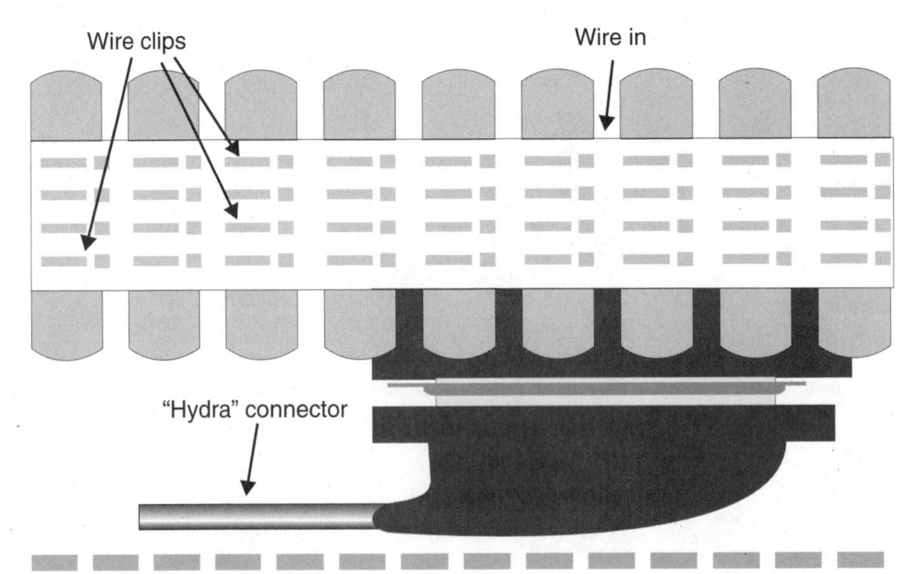

Figure 4.11.
A twisted-pair punchdown panel.

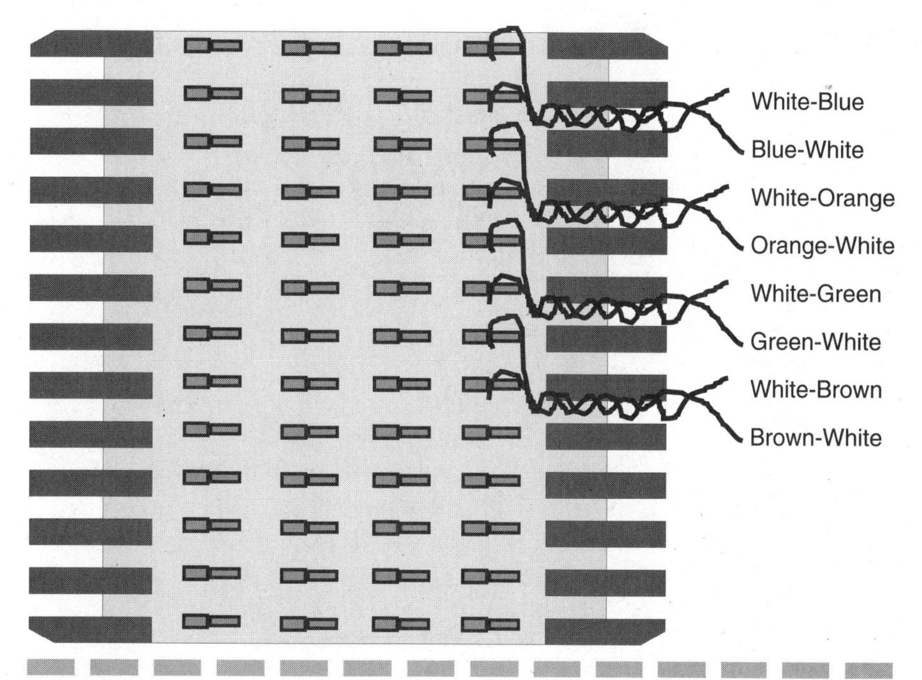

Figure 4.12.
Alternate slots when terminating the block.

down and not operate. This is a good safety feature. The alternative is for the node or hub to assume that the high voltage is a jittering peer. Constant collision signals or requests for retransmission will substantially degrade network performance. Instead, if the single bad element is taken out of service, it is possible that the hub will still function and ignore the failed node.

The wiring closet and patch panel also provide a centralized location for connecting 100Base-T to optical fiber. In organizations with enterprise-wide networking, wiring closets quickly develop into phone maintenance space, computer rooms, and multimedia concentrators and adaptor space. Plan ahead. Provide enough electrical power to support the devices anticipated for this space. Provide enough physical space for the hardware and for people to work. Route cables directly to this wiring closet, or provide enough slack in coaxial cables and optical fiber to later loop them into the room. Consider keeping a live phone in the closet too, because it often takes two people in different places to decipher and repair a wiring problem.

Termination with Connectors

One of the most difficult telecommunications products to terminate is the modular patch cord. Successful patch cord assembly requires considerable skill, patience, and manual dexterity, because each wire must be inserted into the plug or panel individually. This is not like the silver satin ribbon cable that can be inserted into a connector and merely crimped. Proper patch cord termination consists of constructing a product that passes continuity requirements and satisfies the transmission requirements of a specific link. Continuity is an easy check, but there are other requirements for data transmission requirements too.

Connecting points are normally implemented using insulation displacement connectors (IDCs). As of this writing, designs have been verified on an improved version of the original "66" style punch-connect termination block in order to comply with Category 5 requirements. Because each mating connecting point adds one inch of untwisted cable to the overall link, you should not use more than two patch cables per any one horizontal cable segment. Cross-connects are not recommended because of the introduction of four additional connecting points. However, the draft specification allows one cross-connect per installed link. If a cross-connect is necessary, only certified Category 5 cross-connecting cable should be used to maintain link integrity.

In addition to limiting the untwisted cable, it is recommended that the cable sheath be preserved as close to the connecting point as possible to maintain the interpair relationships designed into the cabling. Previous installation practices in which the sheathing was removed from several inches or feet of the cabling severely impair the cable's ability to reliably transmit and receive high-frequency signals. The standard four-pair wiring color code is shown in Figure 4.13.

The actual wiring scheme for modular plugs varies based on the code in force, and whether you are wiring for Ethernet or Token-Ring. I mention Token-Ring here because many organizations have different types of LANs in place, and jumpers and patch cords are not interchangeable. However, if you wire all eight connectors—four pairs as recommended by EIA/TIA 568, rather than the two-pair minimum—straight through, you have a better chance of creating a plug-and-play cable environment. Note that hand-held wiring scanners are typically designed either for Ethernet or Token-Ring, so the cable you qualified with an Ethernet scanner might not pass muster with a Token-Ring scanner. EIA/TIA 568 specifies two different pin assignments. The Universal Service Order

Pair number	Tip or ring	Conductor color
1	T	White/blue
	R	Blue
2	T	White/orange
	R	Orange
3	T	White/green
	R	Green
4	T	White/brown
	R	Brown

Figure 4.13.
Standard 4-pair color code.

Code (USOC) was the forerunner to EIA/TIA standards, and it used a different pin termination assignment for eight position plugs. This is illustrated in Figure 4.14.

These are the actual wiring assignments for 100Base-T4 and 100BaseVG. It is important to recognize that a preexisting Token-Ring only required pairs 1 and 2 to actually be connected, while 10Base-T only required pairs 2 and 3. This points out very clearly the difficulty in migration to 100Base-T4 or 100BaseVG. The actual pin assignments for Ethernet, Fast Ethernet (100Base-TX), and Token-Ring are illustrated in Figure 4.15. If you think that this information seems trite, too technical, or unimportant, just consider that 25-Mbps ATM uses only pins 1 and 2 for transmitting and pins 7 and 8 for receiving, an arrangement very different from those used by Ethernet, Fast Ethernet, four-pair wiring schemes, or Token-Ring.

I mention these variances because it is very practical and insightful to wire premises for any combination of voice, fax, PBX, key systems, Ethernet or Fast Ethernet on UTP, Token-Ring, FDDI over copper, and ATM to the desktop. As you can see, this might require connection of all four pairs to the RJ-45 connector, and a collection of color-coded jumpers to correct for each data connection variation. In any case, correct pair impedance is 90 to 117 Ω, while a split pair would yield impedance between 150 and 170 Ω. USOC wiring is more susceptible to near-end crosstalk (NEXT) because of a severe impedance mismatch. These problems occur because field installers are unaccustomed to USOC. Solution might be as simple as using color-coded EIA/TIA 568A (T568A) or EIA/TIA 568B (T568B) jacks.

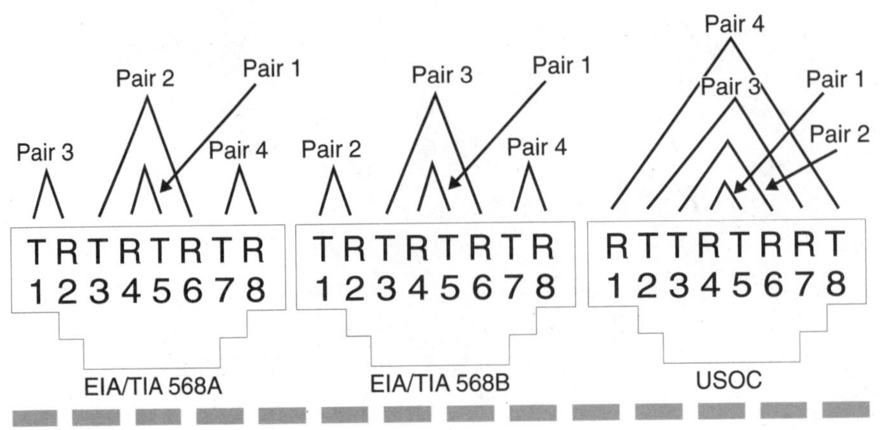

Figure 4.14.
EIA/TIA 568A, EIA/TIA 568B, and USOC modular connections.

The connector requirements are not in themselves sufficient to ensure system performance. You must see to proper cable, cable installation, connectors, and connector installation to achieve Category 3 performance, and pay a lot more attention to these details to achieve Category 5. Some manufacturers have already taken the initiative to develop and implement modular patch cord test methods and pass/fail criteria based on laboratory transmission performance characterization of short lengths of cable and modular plugs. Although the recommendations given here will improve the consistency and overall performance of your field-terminated modular patch cords, it is suggested that Category 5 cords be factory terminated and tested to guarantee proper link transmission performance. Patch cords are also a bear to make by hand, because the fourth wire typically crosses over wires 5 and 6 for tip and ring, as illustrated in Figure 4.16.

Locating Bad Wires and Crimps

Almost any wire can represent a bad connection. This is a common flaw with any network protocol and architecture. It is rare that the wire itself is bad, or will go bad over time. More likely, lobes and jumpers fail at connections, usually at the modular connector itself. The best way to locate these problems is to recognize the error condition indicated by your network management system. In some cases, failure is not outright, but rather a marginal condition indicated by high error rates, frequent requests for retransmission, or the inability for a user to communicate with a server. You might even see a connection failure as a failure to deliver mail, messages, or video. Sporadic and intermittent failure is actually more common than outright failure.

Read the network traffic reports. They are usually indicative of sporadic or intermittent failures. If you can isolate the problem link, you can check status lights on NICs or hubs. Unless you happen to be there at the exact instant a problem occurs, you are unlikely to observe the problem as it occurs. When NICs and hubs include RMON, SNMP, SNMP-2, or some other remote management reporting facilities, enable those services if you haven't, and then review those logs. If you are pressed for time to solve the problem, check lobe and jumper cables sequentially with a handheld wire scanner until you find a bad connection. The better and more thorough devices will test the quality of the entire link through all the jumpers and connections.

If you can isolate the problem lobe, check each wire in the series. If you can isolate the problem to a specific segment, replace it. If you cannot find a clear failure, replace jumpers or repatch the connection. When a jumper or modular connection fails, you are likely to see a bad crimp at a connector.

Such a failure can be constant or intermittent. If intermittent, variations in humidity or temperature could create a sporadic problem. Make a visual inspection. Look from the front of the plug to see if any strands of wire extend beyond the front of the plug. This could cause shorts. Look for proper insulation-áno cuts, nicks, or abrasion within the plug. Check that the main strain relief is actually making contact with

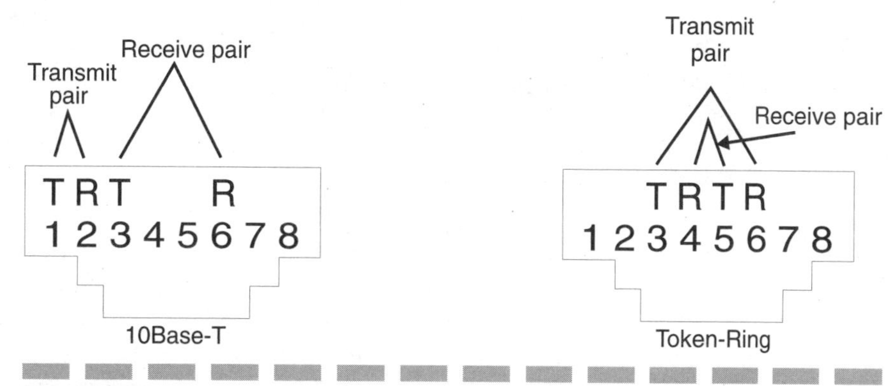

Figure 4.15.
USOC modular wiring comparisons for Ethernet 10Base-T and 802.5 Token-Ring.

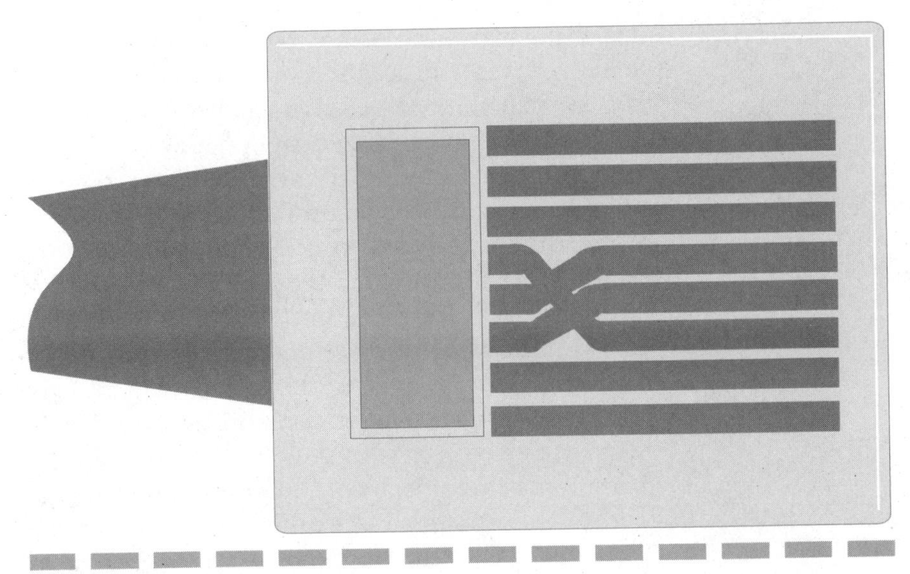

Figure 4.16.
Wire 4 crosses over wires 5 and 6.

the outer jacket. If this is not set properly, check the wire with the wire scanner, and pull on the connector while checking it. Check that this relief is pressing on the outer jacket, not the wires inside. If a secondary strain relief was used, make sure it is seated on top of the wire pairs, not the outer jacket.

Look inside the plug with a magnifying device. Check that all wires reach the contacts. All the wires and contacts should be parallel to each other and not twisted. If poor-quality tools or modular plugs were used, burrs or dirt can prevent proper contact. Also, check that the color code of the wiring corresponds to the pairing chosen for the installation. By the way, you do not need to check unless you experience outright or intermittent failures. Testing lobes and segment wires that are working is likely to introduce new contact failures. It is foolhardy to look for problems where none exist.

These problems are typically caused by time, abuse, or carelessness. When you see these problems with a new installation right from the start, likely causes include poor installation practices, lack of training, casual testing, or poor tools and components. Jacks and plugs are usually not the cause of these problems, although I have seen inferior components create havoc. Often the inferior components are minimal-grade telephony components installed where Category 5-rated components are designated. I've seen a couple of cases where defects in manfacturing created flaws in the materials. Since these components are mass-produced (injection molded and stamped), you will always find a few defects. Do not even consider using suspect parts; at pennies for each part, the cost of fixing a bad component or connection later outweighs any frugality. Because a bad part can create hours of downtime at $1000 to $100,000 lost each hour, do not skimp or cut corners. Also, get high-quality tools, particularly crimping tools. While Radio Shack sells a $2 plier-like crimper, most catalogs and manufacturers sell $40 to $90 tools. These industrial-grade tools withstand abuse and provide the critical wiring measurements required for wire stripping, lead lengths, and exact pressure to pierce the insulation to seat the contacts with the wire conductors. Interchangeable jaws handle silver, satin, and standard UTP or STP RJ-11, RJ-22, and RJ-45 connections. Because Fast Ethernet is unlikely to exist in a vacuum, get a quality tool for other uses.

Jumpers and Patch Cords

Patch cords are assembled from stranded cables of a particular category rating (Category 3, Category 4, or Category 5) to match the rest of your infrastructure. Do not use Category 3 patch cables with a Category 5 installation. However, assembling a cord from Category 5 cable does not necessarily ensure the cord will function properly in a Category 5 link and, because there are no industry standards for qualifying patch cord performance (they're still awaiting final acceptance of TSB 67 and subsequent TR41.8.1 drafts), there really is no such thing currently as a Category 5-compliant modular patch cord. Manufacturers achieve proper cable transmission performance by precisely twisting the cable pairs in special ways, referred to as *cable lay*, to reduce crosstalk. Any disturbance to the cable lay during patch cord assembly might cause link transmission failure.

Begin a modular cord assembly by stripping back approximately 3 cm (1.5 in) of cable jacket, being careful not to nick, twist, damage, or otherwise disturb the conductors. Familiarity with the T568A and T568B wiring schemes will help you to orient the stripped cable and arrange the conductors. These two schemes are different and not compatible. Which scheme you choose might be based on how other work has been previously performed on site, or the standard practice in your area. For example, if your cord is to be wired to the T568A wiring scheme, rotate the stripped cord until pair 3 is on the left. Look at the base of pair 3, where the twisted pair meets the jacket, and verify that the tip and ring of the pair are oriented so that untwisting the pair only to the base will position the pairs properly for the T568A wiring scheme; but do not start untwisting yet. Fan the remaining pairs so that they are arranged left-to-right as follows: pair 3, pair 2, pair 1, and pair 4.

Untwist pair 3 first. Hold the pairs in position by squeezing the base of the cable jacket between your thumb and forefinger. Untwist pair 2 to the jacket base and V the tip and ring conductors around pair 1. Remove twists from pair 1 and orient it between the conductors of pair 2. Position the conductors of pair 4 last. For Category 5 terminations, it is mandatory that the length of untwist does not exceed 0.5 inch. To comply with this requirement, the conductor pairs should not be untwisted beyond the jacket base. You also should not restore cable twisting and add more than there was before the individual pairs were exposed. Remember, cable geometry is critically important to the transmission performance of the final product. Avoid making any unnecessary modifications, including the addition of twists, to the cable pair lay.

When the plug is inserted over the prepared conductor pairs, the conductors should be bottomed out in the front of the plug, and you should see all the conductors when you examine the front of the plug. You can see that every cable will have a segment in its length where all the pair conductors are properly oriented. Cut the cable there. The wires will only need to be guided into position. If you notice that the pairs are not falling into place, simply trim off an extra inch and repeat the process. After the wires are in correct position, hold the jacket in position with your left hand, then grasp the conductors with your right hand and rotate them half a turn clockwise and half a turn counterclockwise. The effect of this is to flatten and straighten the conductors so that they settle in more easily. This will set the wire conductors and make it easier to insert them into the plug.

Next, trim the conductors. The length you trim depends on the plug to which you are terminating. Generally, trimming the conductors so they extend beyond the jacket base by 13 mm (0.5 in) is sufficient. Be careful with your trims; if you remove too much, you will have to start all over. Insert the plug over the prepared conductor pairs. The conductors should be bottomed out in the front of the plug. You should be able to see all the wire conductors when you examine the front of the plug. Push the cable jacket past the primary strain relief of the plug. Then crimp the plug and you are done. When you are done, the whole thing reduces to Figure 4.17.

Typically, premise wiring requires one more layer of jumper between the punchdown panel and a separate patch panel with modular connec-

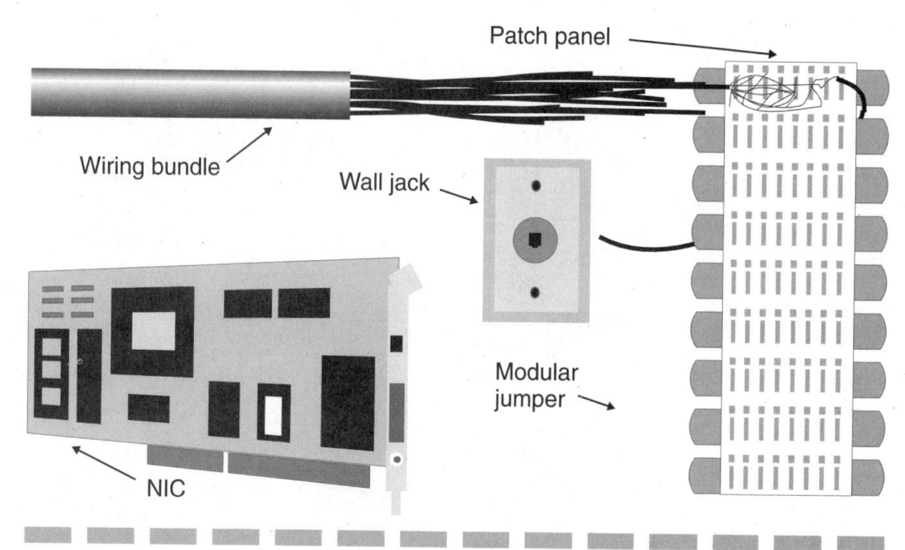

Figure 4.17.
Twisted-pair wiring for plug and jack.

tors. The hydra or handfuls of wires are connected from the punch-down block to back of the modular panel. This cross-connect facility increases flexibility for future wiring changes.

Near-end/Far-end Crosstalk

Crosstalk is the result of a signal in a pair of wires infiltrating another pair. Crosstalk is a major wire failure problem with Fast Ethernet. Excessive crosstalk usually indicates substandard components or improper installation. Crosstalk also occurs when lobes exceed length limitations. Crosstalk is more likely to be a problem with 100Base-T because of the higher transmission frequencies, but this assertion should not be considered to discredit 100Base-T.

When you find one lobe with excessive crosstalk, you should check more. If the problem does not appear when a wiring infrastructure is first installed but rather develops over time, check your change log to narrow the search. You do have a change log? If you can narrow the failure to a connection, it is possible that excessive wear or stress wore out the component. Also, realize that crosstalk with Category 3 and Category 5 wiring is defined for only 20°C (68°F). Because wiring often crosses piping, lighting fixtures, runs through air-return plenum, that temperature measurement is unlikely to reflect the entire cable run. Heat and humidity increase the crosstalk. Factor the reality into your testing. Also, maintain logs of past test results.

Typically crosstalk occurs at connectors, where the cable ends. This is not the rationale for why crosstalk is termed *near-end* and *far-end*. Rather, this is a function of how the crosstalk is measured. Of course, you must connect a scanner into the wire from one end or the other, and this is where the terminology comes from. Although tools exist for confirming the presense of signal or voltage inside wires, you still need to attach a measurement device from one end or the other to test crosstalk. Figure 4.18 illustrates testing for crosstalk from the work area with a handheld tester.

The readings on any particular cable run typically vary from one end to the other. The differences are immaterial for capacitance, resistance, and impedance, but not for crosstalk. If you have access to a typical pair scanner, you will also want to test the far end at the wiring closet for crosstalk problems as well. This is illustrated in Figure 4.19.

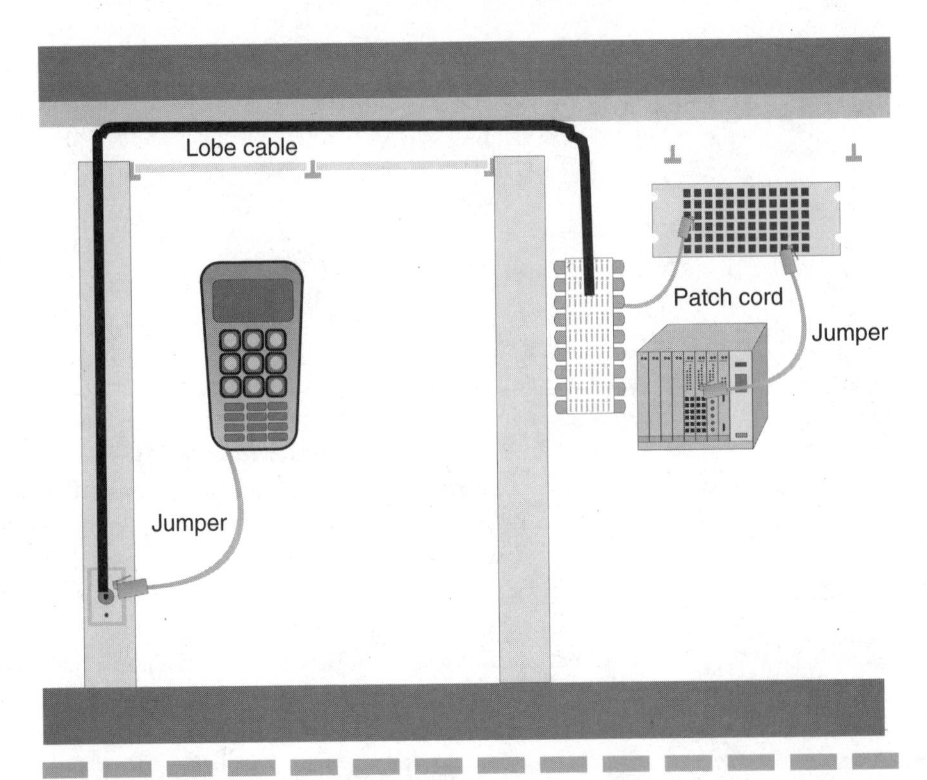

Figure 4.18.
Testing for crosstalk with a handheld tester.

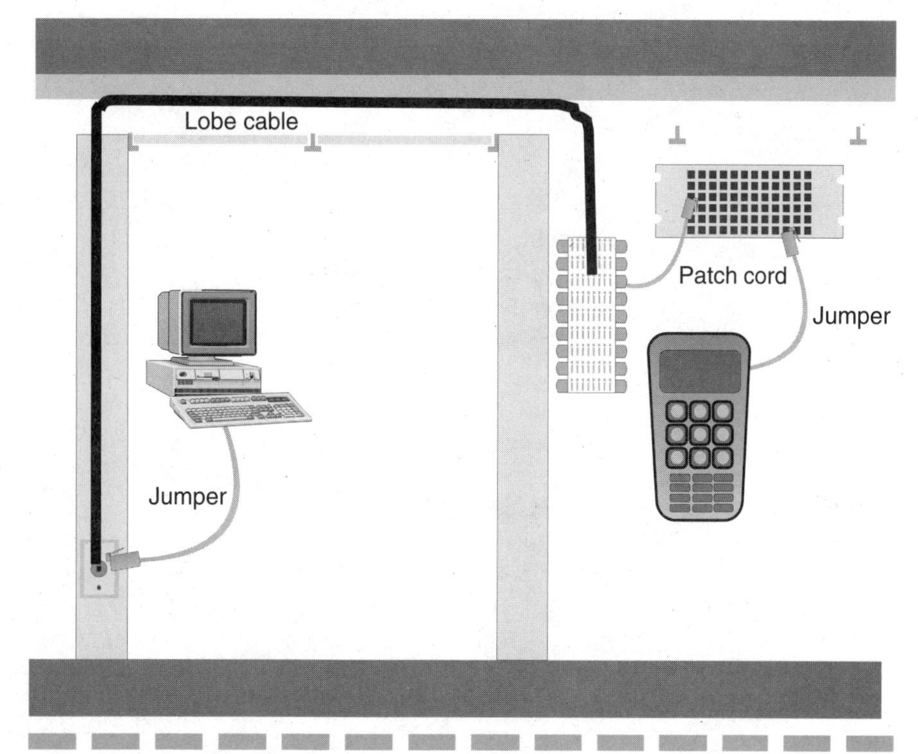

Figure 4.19.
Testing for the far-end crosstalk with a handheld tester.

Older pair scanners designed for 10Base-T or Token-Ring perform at 10 to 16 Mbps. Some of these devices also presciently support 100 Mbps. These tools often require a helper with a phone, cell phone, or walkie talkie to coordinate the end-point pair testing. It is very tiring doing this yourself in a big building, because you end up walking back and forth for hours. Basically, you need to terminate each office and test from the wiring closet for the crosstalk levels. This is illustrated in Figure 4.20.

Incidentally, the illustration shows the scanner being attached from the office. I think that, from a workflow standpoint, this is backwards, although this works just fine and which end you test does not matter with these two-part tools. However, because most wiring is not labeled well, it is easier if the helper inserts the loop-back terminator into the jacks in each office. You test the pair from the closet and label it as you go. If the ends do match up, you have more possibilities to locate the correct set of pairs from the wiring closet.

Ceiling Attachment

Fire codes usually require a false ceiling in all parts of a building. Wiring closets and utility are rarely immune from these rules. However, because computer rooms often grow inside a closet or an office space, sometimes you will need to retrofit the wiring to meet code. Replace cut and notched ceiling tiles with new ones. Cut a 3.75×2-inch rectangle in the center (or make a cardboard template from a ceiling faceplate) and insert the faceplate. Terminate wires above the faceplate and label. Create jumpers between devices in the computer room and that faceplate. This is illustrated in Figure 4.21.

Optical Fiber

Optical fiber is available in more variations for FDDI than are typically used for 10Base-FX or 100Base-FX. Fiber varies by length, thickness,

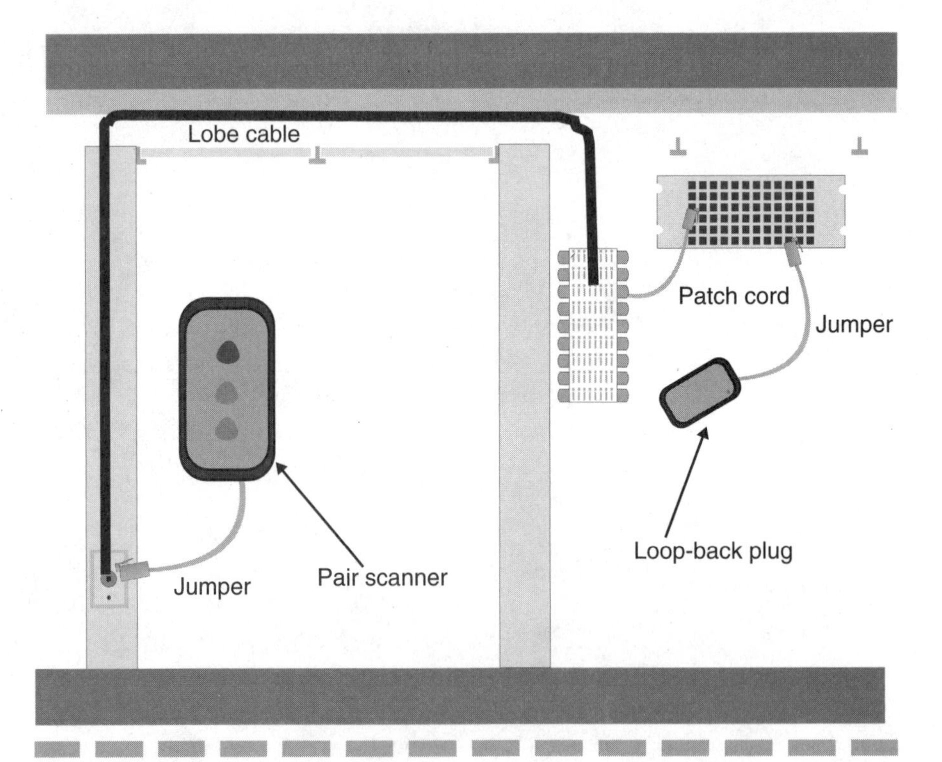

Figure 4.20.
Two-part testing for crosstalk with older equipment.

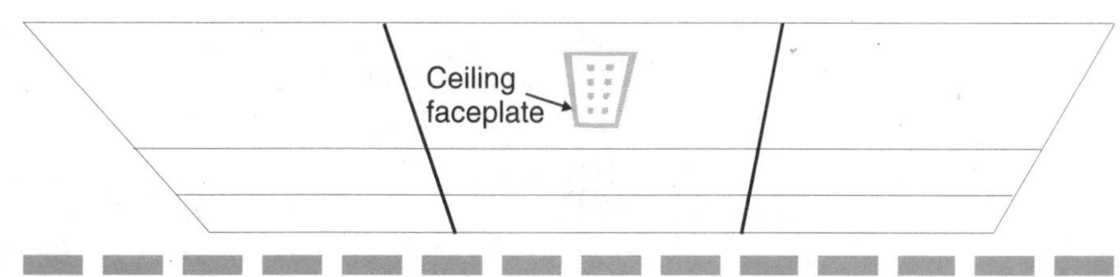

Figure 4.21.
A T-bar ceiling wiring faceplate.

packaging, and modality. The standard and recommended optical medium is 62.5/125-mm fiber (single-mode at 850 nm signaling), packaged as fibers for point-to-point Ethernet links. You need a fiber for each direction. If you buy fiber cut to length, buy it with ST connectors installed and tested. By the way, the ST connector is very similar to the BNC connectors used in Thinnet, only smaller because the fiber itself is smaller (and lighter). If a vendor is installing the fiber, they will install the connectors on site and then test them. Other fiber is delivered in spools. Cut it with clippers to length. Polish it and then attach connectors with the specialized crimping tool, epoxy, or by whatever method is indicated by the connector vendor.

Optical fiber is available in various thicknesses. The thickness measurements are paired. The first measurement refers to the internal transmission-carrying fiber; the second refers to the outer diameter. The measurements are important because connectors must match the sizes to fit properly and securely. Typically, fiber used in Ethernet is 62.5/125 mm, whereas other sizes are 50/100 mm and 100/125 mm. The smaller the internal dimension, the better the fiber signal transmission properties provided by the fiber. In other words, the connections and polishing can be sloppier with 50/100-mm fiber, there can be more splices and jumpers, or the point-to-point connection can be longer. Conversely, the military typically specifies 100/125-mm fiber so that troop training time and the time to repair connections in the field are minimized.

Modality refers to the signal "color" transmitted by the fiber. Different wavelengths of light can be transmitted on multimode fiber to carry simultaneous or spread-spectrum transmission channels. Single mode supports only one wavelength and one channel. Your need to verify it with your fiber equipment vendor, but Ethernet on fiber usually requires multimode fiber. Calibrated wavelengths include 850, 1300, and 1500 nm. 100Base-FX is almost always 850 nm with an 850-nm LED signal source. Two fibers (dual cable) of the same type can also be used instead of the pair bundle. However, the difficulties involved in telling which is the primary and which is the secondary usually push veterans to choose the simpler dual-fiber packaging. Optical fiber can be obtained with multiple pairs (20 pairs or more) in the same bundle. It can also be packaged with CATV cable, telephone UTP, and data-grade or STP cable. Manufacturers will also fill special orders.

Optical fiber wiring is typically point-to-point; one station connects to the next station. Jumpers are rarely used, in order to minimize the number of connections. Each connection reduces signal strength; each physical connection causes a –1.5-dB loss or greater. A splice causes a 0.25-dB loss or greater. These vary by the connector, the connector method, and the skill of the installer.

When pulling fiber over the distances usually required, you will almost always require a pulling lubricant to speed installation, lessen the stress on the cable, and negotiate bends and corners. Water is effective and free, but some installers prefer a polysilicon or Teflon gel and intermediate pull boxes for snake access. In some cases, underground conduit or even conduit in a plenum might contain trapped stagnant water. Run a garden hose into the conduit, attach it (with mating connectors) to a shop vacuum cleaner, and remove the water. Figure 4.22 shows how to drain a conduit with a canister vacuum.

You can also use the shop vacuum to blow a parachute attached to a pull string through the conduit. A parachute is more forgiving of bumps and bends in a conduit than a long snake. Although some cable might be stiff enough to stuff down the conduit, you usually get better results by pulling the cable instead. Figure 4.23 shows how to use a paper plug or plastic bag as a parachute.

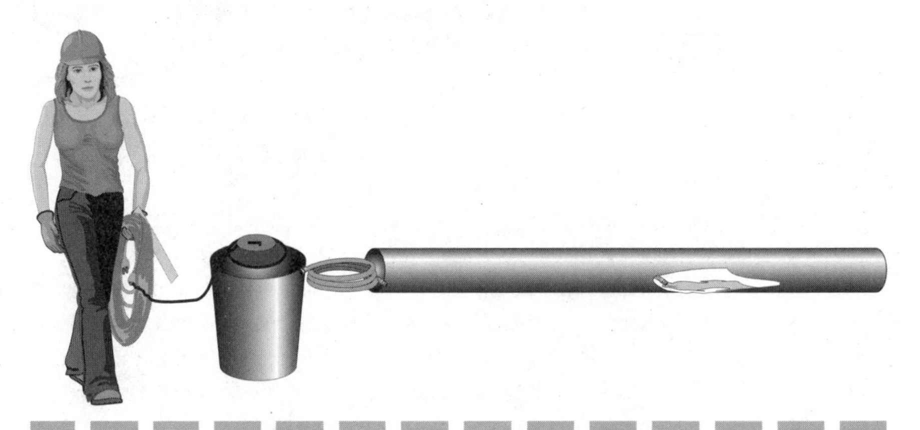

Figure 4.22.
Use a shop vacuum and hose to drain water from conduit.

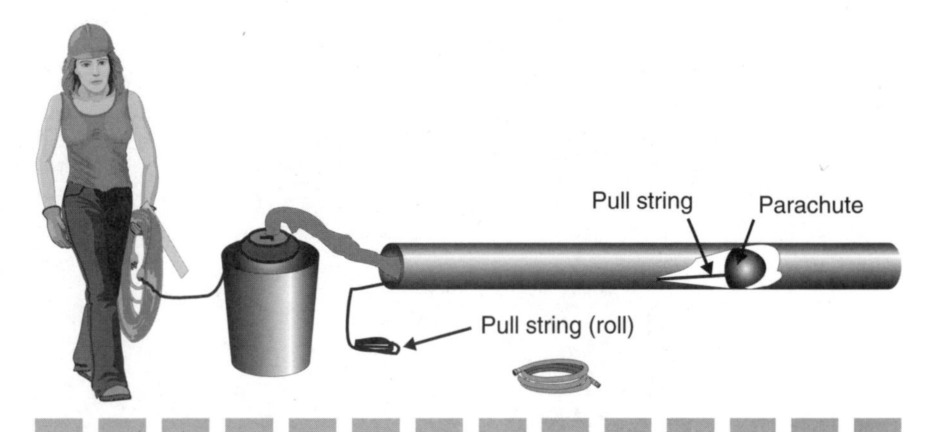

Figure 4.23.
A shop vacuum can blow a parachute with pull string through an accessible conduit.

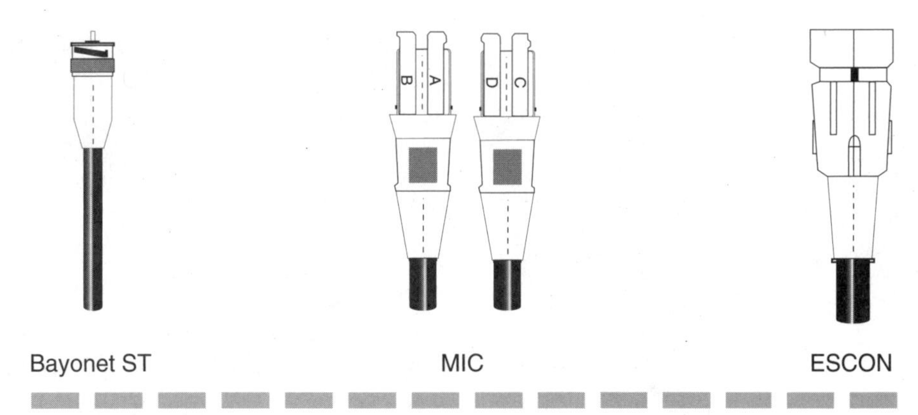

Figure 4.24.
Typical optical fiber connectors. Fast Ethernet usually uses ST, although the norm for FDDI is MIC.

Single bayonet connectors are used for attaching test leads and the repeater link stations. Duplex connectors are more common, and are the preferred method for FDDI; the key does not fit the single bayonet connector most often used. Figure 4.24 illustrates the various optical fiber connectors. Most FDDI installations will use a keyed variation on the MIC connector, while Ethernet on FOIRL typically uses a paired bayonet ST assembly for "ring-in" and "ring-out."

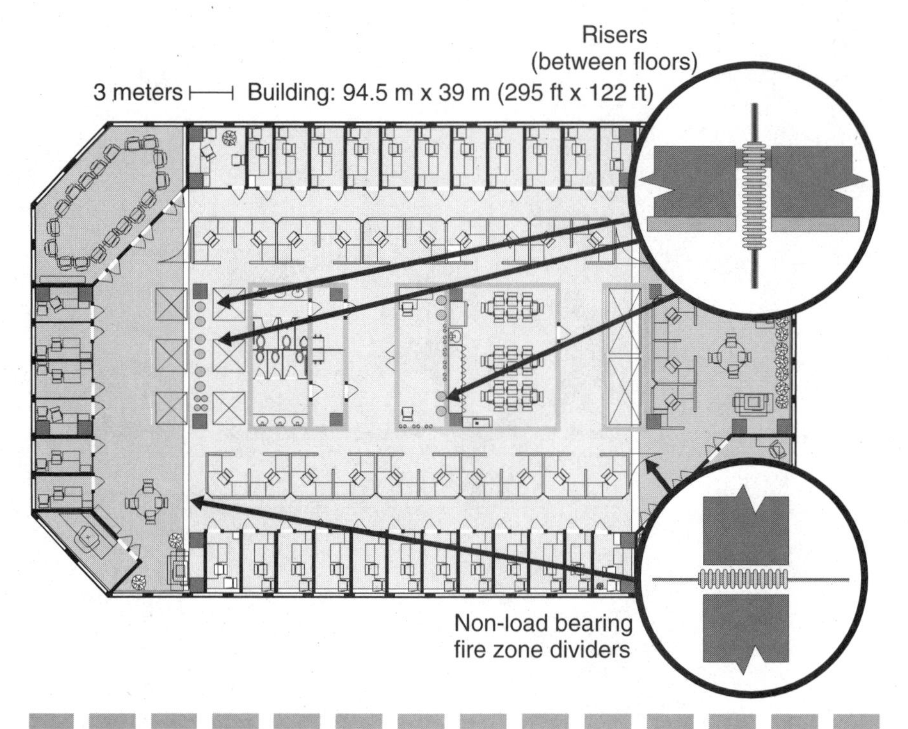

Figure 4.25.
Location of holes in the fire containment areas.

Fire Safety

Whenever cables and wires penetrate fire walls or pass between floors, it is usually legally necessary and ordinarily advisable to install fire-stops. Even when risers are installed into large-diameter conduits, the ends of those conduits often require fire stopping. Fire stops must be installed wherever a fire-rated floor or vertical firewall is penetrated. The standards for this include ASTM-E119, ASTM-E814 two-hour "F" and "T" ratings, and UL-1479. The blueprint (Figure 4.25) shows vertical and horizontal holes breaching the fire containment areas, and the types of fireplugs required.

Fire stops usually are fire-resistant silicon-type gels, sleeves, saddles, foams, or putties used to fill gaps around cables in conduits and holes, as Figure 4.26 shows.

Fire-stop gels and caulk are typically available as cartridges and installed with an ordinary caulking gun. Firestopping pillows are for temporary use only. They are compressible packets of what is basically a cementlike material; it is called refractory material, because it doesn't burn. Vertical holes between floors are the most serious breaches to consider, because when a fire occurs, natural convection causes the heat to rise. Holes drilled or punched through concrete walls are dangerous too, as even a small hole allows fires to spread. You might need to employ the building architect or a qualified consulting architect to read the building plans, explain the local building codes, and walk the building with you. Have the architect point out where fire containment cells have been compromised by retrofitted air ducts and plenums, cable trays, and wiring paths, so that you can install the necessary firestopping caulk or firestopping pillows.

The essential danger is that you do not want fire to penetrate outwards and particularly upwards to floors above, a problem best depicted in the movie *The Towering Inferno*. When a fire occurs, the combustion temperature easily exceeds 1200°C within five minutes. The extreme heat vaporizes many materials, including paper, wood, plastic, and even the copper in network cabling. In this fatal environment, the now-gaseous material cannot burn, because all the available oxygen within the fire area has been consumed by the fire itself. Combustion gases in this situation are oxygen-starved and cannot burn. If these superheated gases infiltrate an

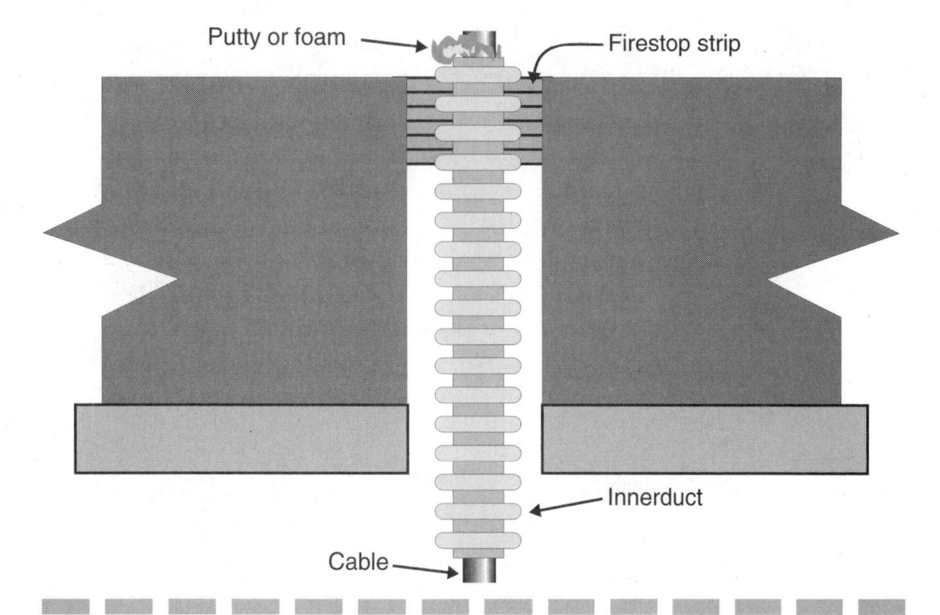

Figure 4.26.
Firestopping wrap strip and moldable putty fills holes for conduits through firewalls and vertical risers through floors.

upper floor, the temperature of these gases alone is sufficient to cause paper, wood, Formica, pressed-board furniture, and carpeting to ignite. However, because these gases are themselves highly incendiary, the primary danger is really that gases will spontaneously ignite in the presence of the fresh air in the new area. Ignition is immediate and explosive. Imagine spraying superheated gasoline into the air; it will explode. Even small cracks are sufficient to allow gases, given the vapor pressures of the fire, to penetrate upward and outward, spreading an otherwise confineable fire beyond the fire cells designed into modern structures.

Architecture

The 100Base-T specification is a star configuration, and hubs interconnect as either workstations with network interface cards (NICs) or hubs. 100Base-T service requires supporting equipment for hubs and NICs. Some vendors provide hubs that "autosense" Ethernet transmission speed; I do not recommend that you mix 100Base-T and 10Base-T at a hub, because the great differences in speed can create a bottleneck at that hub because of translation-induced latencies.

Note that some vendors call the hub a multiport repeater, a wiring concentrator, or a wiring center. The hubs are available as stand-alone hubs, stackable hubs with an AUI or backplane attachment for daisy-chaining, or as chassis-based hubs with gigabit backplanes and interchangeable slots for 10Base-T, Token-Ring, 100Base-T, FDDI, FOIRL, and network management station modules. The signaling on the 100Base-T, 100Base-T4, and 100BaseVG is a phase-locked loop circuitry that detects a packet preamble; this is true regardless of signal coding methods. This starts the clock for packet capture. The hub should provides all packet retiming and retransmission, filtering, and timing signals.

A few problems need to be solved for Ethernet to work on UTP. The cabling must not be too susceptible to electromagnetic noise (from transmitting devices, fluorescent lights, etc.), other devices transmitting frequencies near to that of the 100Base-T signal, crosstalk between twisted pairs, and jitter (which occurs when preamble or packet signals are out of phase). 100Base-T performance is a function of the signal-to-noise ratio. Higher-quality cable, better cable installation, shielded cable, low utilization of all the pairs in the cabling bundles, superior wiring connections, shorter wiring runs, and superior hardware (including hub concentrators and the controller/transceiver cards) improve the chance that the signal will reach its destination. It is important to remember that Ethernet is a chance delivery; the management game is to improve that chance.

Existing 10Base5 and 10Base2 installations can be converted or interconnected to 10Base-T with microMAUs, and then wired into a Fast Ethernet. Those slow segments will not benefit from the rest of the transition to 100 Mbps, but can get physical network access at the slower speed. You do not want to exclude network access, but integrate it; this is how you do it. This microMAU transceiver replacement converts the phased-locked transmission into a format consistent with that required by 10Base2 and 10Base5. You can convert either way, from 10Base5 to 10Base-T or the reverse. While some microMAU units correct for polarity errors inherent in voice-grade UTP wiring, you are better off correcting and qualifying the wiring infrastructure. A typical unit—this is an OSI Layer 1 media converter or repeater—is illustrated in Figure 4.27.

Twisted-Pair Wiring Tips

Twisted-pair wiring is ubiquitous in office buildings, it is cheap and an old technology, and it fits the concept of the star topology for ease of installation and network maintenance. Shielding is nonexistent, routing is typically suspect and not documented, and the telephone cable probably does not meet the more rigorous specifications required by data connections. It rarely will conform to EIA/TIA Category 3 and (as you have seen) might be wired for USOC rather than EIA/TIA standards. If you want your network to be reliable, the road to twisted-pair networking should begin with a new cable job. It is important to differentiate

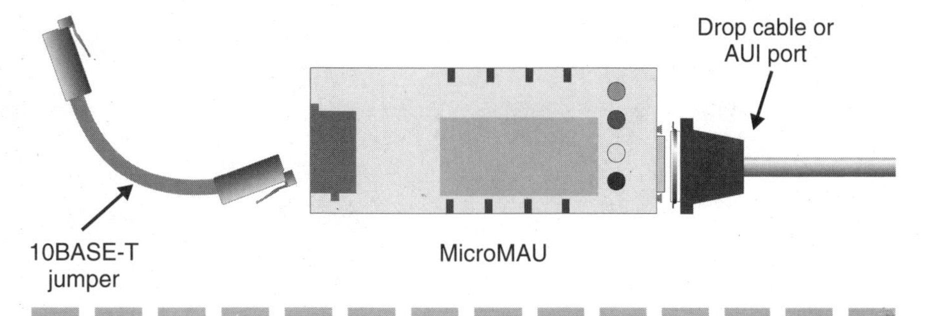

Figure 4.27.
MicroMAU converts LAN cabling from coax to UTP for migration and interconnectivity of Ethernet to Fast Ethernet.

unshielded twisted-pair from shielded twisted-pair cabling. Poor wire could increase jitter. Twisted-pair wiring installation tips include:

- Ensure that straight-though polarity is maintained.
- Use cabling that meets operating specifications.
- Use cabling that meets fire code specifications.
- Maintain a uniform color code.
- Maintain 100-Ω impedance.
- Limit the use of flat cable to under five feet on any connection.
- Avoid shielded cable with baluns, multiplexers, or patch panels.
- Use data-qualified punchdown panels.
- Route twisted-pair away from RF and electromagnetic radiation.
  - 6 inches from standard power lines
  - 12 inches from fluorescent lighting
  - 36 inches from transformers
- Check for loose connections.
- Use RJ-45 connectors (not RJ-11).
- Observe pair polarity.
- Do not split pairs.
- Apply a conductive spray to connections when they're installed.
- Hire a qualified data cable installer transmission.

You can estimate lengths of existing wire by measuring dc resistance with a multimeter (or use a pair scanner with a mapping function). For example, 1000 feet of 24-gauge wire will measure 16 ohms. It is important to note that RJ-11 plugs and jacks are physically interchangeable with the RJ-45 counterparts. However, the wires will not match up correctly. Be certain that the cable installer recognizes that there is indeed a difference and installs the RJ-45 connectors for data lines. Consider acquiring a large supply of data-grade RJ-45 (Category 5) components in a nonstandard color to provide immediate and effective visual differentiation. Also, avoid intermixing the standard telephone punchdown panels for two reasons. They are not reliable for data connections, and they encourage incorrect wiring and possible cross-connections with active telephone circuitry. This causes hubs to shift into a link failure mode and not transmit on the faulty lines.

Modular Furniture

Cubicles and modular furniture is an issue for Fast Ethernet wiring. There are three issues of importance:

- Coresident power and data lines
- Tight bending radius within and around built-in conduits
- Proximity to metal

In general, Fast Ethernet will fail when wired within modular furniture and partitions because power is built in, the surrounding structure is mostly metal, and the bend radii are tighter than 5 cm (1.5 in). However, you can drop cables from the ceiling to partitions and modular furniture successfully. Minimize proximity to power and metal conduits. EIA/TIA and the manufactures of conduits and modular furniture are trying to define what will work so that they can design components that will work with high-speed networks, and potentially will include a built-in wiring system for power, phones, and data.

Cable Testing

Most new Fast Ethernet installations specify that the physical layer (cables and connectors) be capable of transmission performance up to 100 MHz. The specific standard usually cited for purposes of certifying the cable system is, of course, EIA/TIA568 Category 5. Ironically, although there are many Category 5-compliant components available from which to build the physical layer system, EIA/TIA 568 contains no specification that defines the criteria for certifying the resulting cable infrastructure for Category 5. SP-2840 is the first attempt to define certification criteria for the physical-layer components connected as a system. Because this specification has not been ratified, it cannot be said to define pass/fail criteria for Category 5 certification. However, it addresses the important technical issues and provides the best guidance available as to the kinds of tests that must be performed to ensure the integrity of a high-speed LAN installation that is designed to be "Category 5."

The proposed testing criteria in SP-2840 is complicated. Installing a Category 5 cable system is not simple. Even when properly compliant components are used, many errors can occur in the very act of putting the pieces together, such as kinks, poor connections, and excessive lengths. Many installation practices that work well for lower-performance links cause problems when used to install systems that must perform at speeds approaching 100 Mb per second. This is true because these signal strengths are high enough that even small imperfections in the system can result in increased crosstalk and attenuation.

It is important to minimize crosstalk and attenuation in Fast Ethernet because the LAN segment provides two data-carrying circuits, one to carry data from a hub or central data server to a workstation or peripheral, and the other to carry data in the reverse direction. The close proximity of the two circuits means that any signal radiated by one will be received by the other and combined with its own signal; i.e., crosstalk will occur. But in order for the receiver at the end of a circuit to recognize the incoming message, the signal must be stronger than the amount of induced noise from the crosstalk; i.e., a high signal-to-noise ratio must be maintained. Thus, the ultimate determinant of whether a cable system will work is the difference between the measured near-end crosstalk and the measured attenuation. The higher the frequency, the smaller the worst-case difference and the greater a chance for a transmission problem.

Ensure connectivity for all four pairs to meet any topology. After connectivity, the most important tests are for the link's attenuation and near-end crosstalk characteristics. Cable system attenuation or loss should be tested one pair at a time using a sweeping signal source in conjunction with the handheld cable system tester. Loop-back style testers are affected by the performance of another cable pair. Figure 4.28 shows a typical handheld unit.

As frequencies increase, it is important to differentiate which pair is providing what level of performance. Because all measurements of attenuation are referenced to a 0-dB source signal, the smaller the measured attenuation (in dBs), the better. For example, a -4.5-dB measurement at 100 MHz would be better than a -6.5-dB measurement at the same frequency. Near-end crosstalk should be measured between all pair combinations for complete Category 5 compliance. NEXT behavior is very unpredictable, and must be tested using a sweeping source with measurement increments of no more than 2 MHz. A typical NEXT test from 100 kHz to 100 MHz will include 301 stepwise measurement points. Unlike attenuation, which gradually and regularly increases with test frequency, NEXT exhibits peaks and valleys of performance throughout the measured frequency band. A sweeping signal source from a signal injector in conjunction with the handheld tester is the

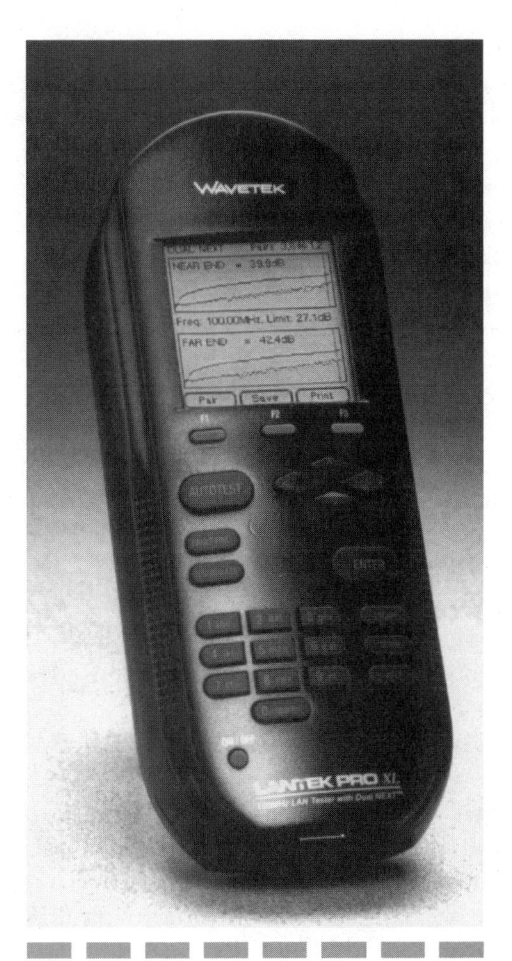

Figure 4.28.
A handheld wire tester. (*Wavetek*)

most reliable for testing. Remember, you must measure near-end crosstalk at 100 Mhz to be certain of the results at that level. It is impossible to extrapolate the results by projecting a curve based on a measurement at a lower level, such as 20 MHz.

Crosstalk is usually measured at the near end because that is where the transmitted signal and the potential for NEXT are the strongest. Signals induced from adjacent lines can have sufficient amplitude to corrupt the integrity of transmitted data and interfere with reception of the data. Near-end crosstalk is one of the most important tests to perform, because excessive NEXT can cause problems ranging from poor LAN performance to complete network attachment failures. Not long ago, LAN transmissions of 10 Mbps were typical. However, today's Category 5 installations have dramatically increased the number of LANs with transmission frequencies approaching the 100-Mbps range. Because crosstalk increases as the signaling frequency increases, controlling NEXT has become a particular concern to those charged with installing and maintaining high-speed networks such as ATM and Fast Ethernet.

Copper cabling is often installed during new construction because the open walls, floors, and plenums provide simplified access to cable trays, ducts, and risers. Because partitions and the niceties of an office environment might not be added until weeks or months later, basic link testing only confirms the integrity of the cable pulls, not the connections; there aren't any connections yet. Not only should the communication wiring be checked for basic integrity, but the connectors and jumpers need to be reviewed after they have been installed. Typically, the data network is checked out at the same time the PBX and services are installed. However, the EMI of a building under construction can be very different from that of a building ready for occupancy.

With all lights, the heating system, the ventilation system, the cooling systems, and any unusual devices all turned on, check the system. You are likely to discover wiring routes disrupted by elevator motors, fluorescent lights, bright halogen lights, or a kitchen with its microwave oven and refrigerator. Electronic noise might not have been a problem with 10Base-T, but it is likely to become one with any 100-Mbps Ethernet, and especially with 100Base-T (because the line signal is broadcast at 125 Mhz). Intermittent or seasonal problems might be indicative of a heating or cooling motor. Typical EMI sources are represented by Figure 4.29.

Testing cable that is near metallic surfaces might cause an additional 3-percent increase in attenuation (as expressed in dB). Metallic surfaces include ductwork, T-bars for hanging ceilings, structural steel columns and beams, and other cabling, pipes, or power lines. A change in relative humidity from the typical 40 percent to 90 percent might cause an additional 2-percent increase in attenuation. Changes in temperature might be reflected in varying levels of humidity throughout the length of the cable run, and both changes are temporary and reversible.

The best way to locate EMI problems is by tracing cable routes and looking for possible sources. Shut off the possible sources one by one and see if that solves the interference. The operational solution is to move the source, move the high-speed cabling and reroute it, or replace it with a shielded cable. By the way, if you opt for STP, do not just tape an STP segment to the unusable UTP run and pull that UTP back

EMI Type	Frequency	Source
Low frequency	10 to 100 khz	Electric heaters
		Fluorescent lights
Medium frequency	100 kHz to 1 MHz	Radio
		Networks
High frequency	1 MHz to 1000 MHz	Radio
		TV
		Computers
		Computer networks
		Motion sensors
		Radar
Random impulses	10 kHz to 100 MHz	Motors
		Switches
		Automobile ignitions
		Welding equipment
		Electrostatic painting
		Lightning

Figure 4.29.
Sources of electrical interference.

through the run; the wire should be tied at various points, and the replacement should be fastened down, too.

After the transceivers or wiring and hubs have been installed, test the installation. It's a good checkpoint in the installation process, because the network is now functionally complete. Correct wiring and some common problems are illustrated below. Figure 4.30 shows correct wiring. Contrast this with Figure 4.31 showing a wiring break, the short in Figure 4.32, and excessive untwisting of pairs in Figure 4.33.

If the network is broken, disassemble it and restart the process. The most common reasons for network failure are careless installation of a connector, shoddy installation of a cable, and cross-connection of a live phone circuit with UTP. Twisted-pair should be tested for correct polarity and no crossover of the pairs. While many NIC and wiring concentrators do switch contacts automatically in instances of polarity exchanges for pairs, they cannot handle miswiring of both pairs. You

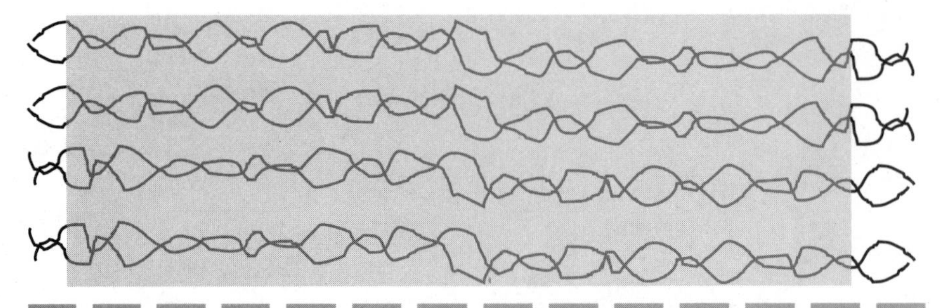

Figure 4.30.
Correct wiring—straight through.

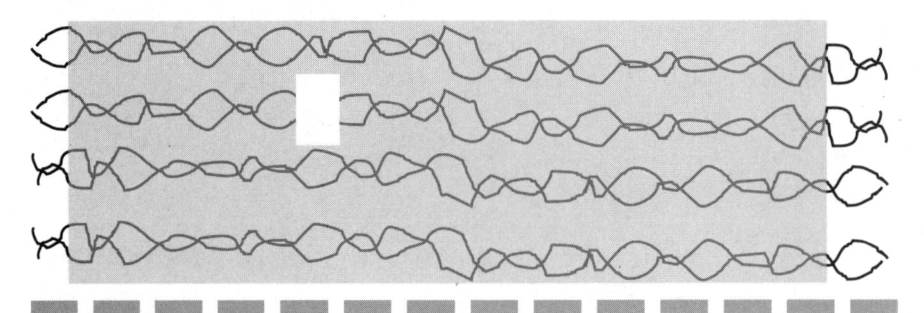

Figure 4.31.
A break in the wiring.

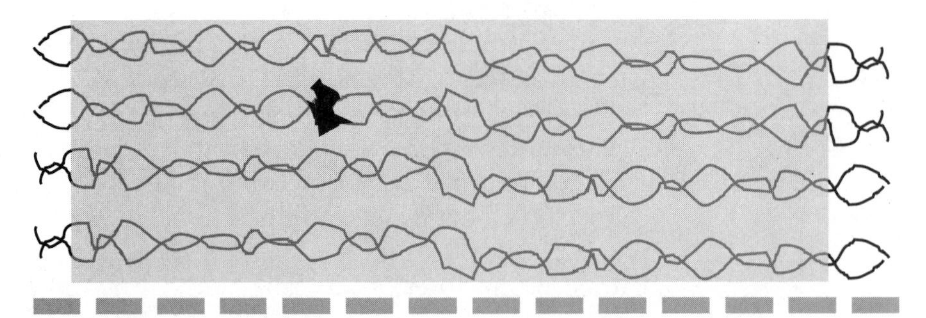

Figure 4.32.
A short in the wiring.

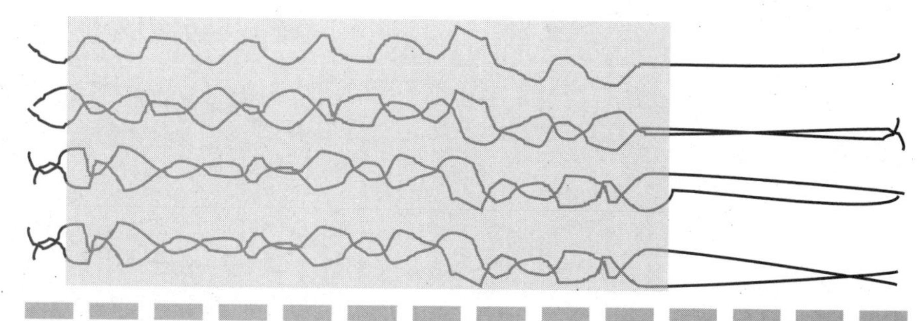

Figure 4.33.
Excessive untwisting of pairs leads to NEXT.

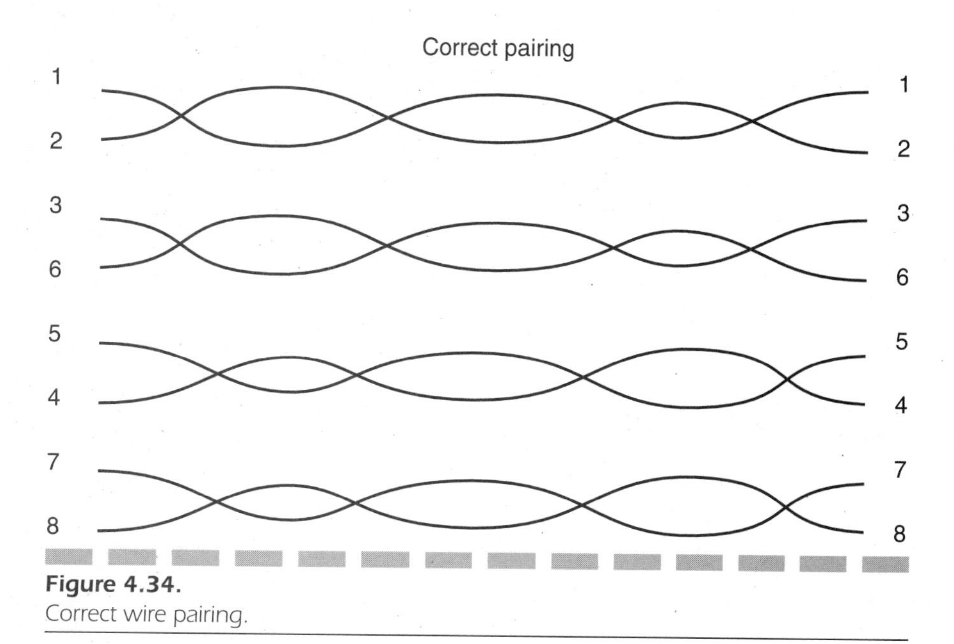

Figure 4.34.
Correct wire pairing.

could have none of the wiring problems displayed in the prior four illustrations. These are uncovered with only an ohmmeter, but you could have these next problems, which are best uncovered by a more sophisticated pair scanner. Figure 4.34 shows the correct pairing.

Figure 4.35 illustrates reversed wire pairing, Figure 4.36 shows transposed wire pairing, and Figure 4.37 shows split wire pairing. These problems really are irrelevant for 10Base-T, but represent a serious flaw for any Fast Ethernet.

Series inductance (evident as a scrape or breakthrough the UTP wire shield) shows as a positive pulse followed by a negative pulse. The last

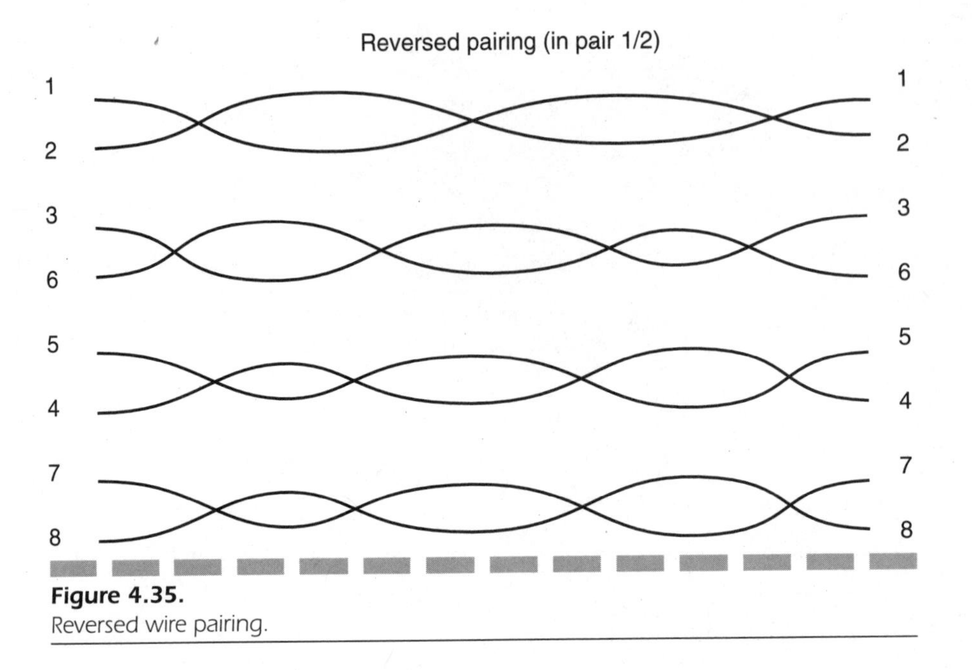

Figure 4.35.
Reversed wire pairing.

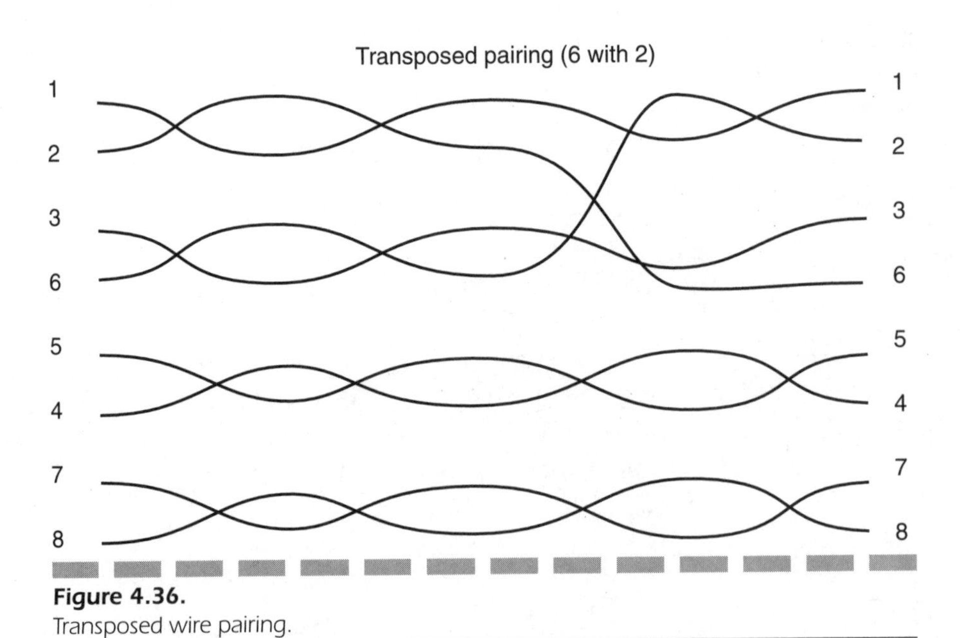

Figure 4.36.
Transposed wire pairing.

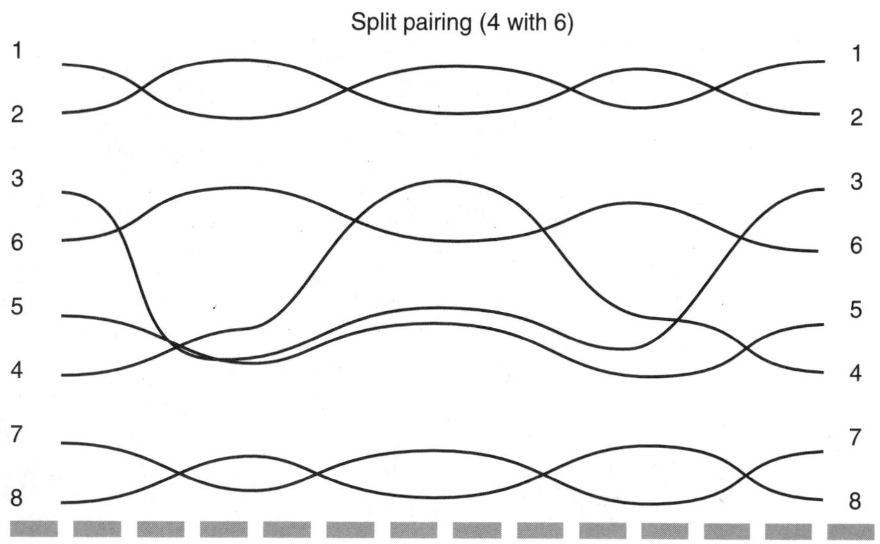

Figure 4.37.
Split wire pairing.

common waveform shows that the network cable impedance is out of specification, as would happen when improper cable is installed. This is typical when voice-grade UTP of insufficient quality is applied for lobes or runs longer than 30 to 50 meters. A series of alternating pulses would suggest this problem. Figure 4.38 details a scanner image caused by spliced twisted-pair wire with different transmission characteristics. You should install a single wire for a lobe and limit jumpers to two or at most three.

Breaks are more common than shorts. Pairs that cross over will break the network, a live telephone circuit in place of the proper LAN lobe certainly breaks the circle, and partial media connections or failed punch-down ties will halt the network. While node beacons might provide a general indication of where the break is, the scanner is the superior tool for actually locating and fixing short circuits, open circuits, or breaks. These tools supersede the multitester because they accurately position the problem. There are also certain problems that cannot be diagnosed by just a multitester with any immediacy. These include bad wire, the wrong grade of wire, radio-frequency interference, internal flaws, and mismatched sections. While a multitester will identify that one of these problems exists, it will not pinpoint the problem so that it can be categorized, localized, and diagnosed. Once you know what the failure is, review the table in Figure 4.39 for possible causes of the failures.

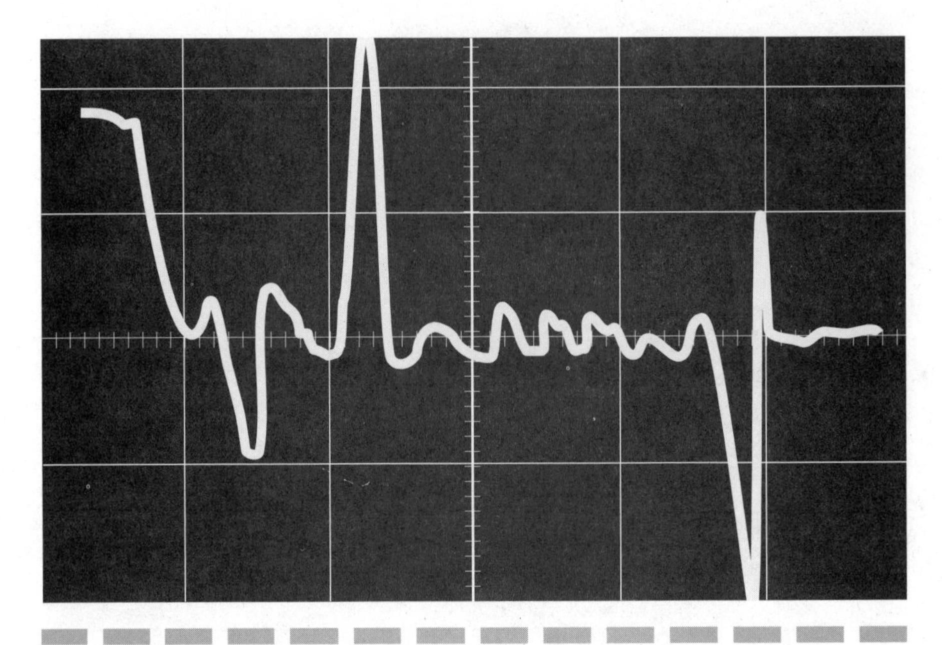

Figure 4.38.
A screen shot from a cable tester.

Optical Fiber Testing

The primary tool for testing optical fiber is the optical time domain reflectometer (OTDR). However, you can use the new breed of handheld pair scanners with an accessory kit to turn the units into functioning OTDR units at a fraction of the cost of the handheld unit, and larger fraction of the cost of a complete stand-alone OTDR unit. One such accessory kit is shown in Figure 4.40.

This is how copper and coaxial cables were tested before handheld pair or wire scanner were available. A typical OTDR bench-type unit contains an oscilloscope display, a TV-like screen that displays a trace curve on a grid. The technique applied by such a tool was developed in the early 1950s for locating cabling breaks in high-tension wires.

The TDR was further refined specifically for use in tracing cable television broadband problems. It was used to locate shorts, breaks, cable impedance variances, and overloads on city-wide installations, but has been modified for use in baseband and broadband computer communication environments; the problems encountered are identical. The same technology was converted to light-based testing with the first fiber-optic networks, and the technology hasn't changed very much. However, just as there are signal injectors for copper wire testing, there are light injectors for fiber. Also, there are other tools such as fiber talk sets, light meters, and loop-back accessories suitable for testing the FDDI or 100Base-FX links.

The OTDR unit connects onto the network in place of a node. The standard OTDR or pair scanner will not send out a phantom signal and

Measurement Result	Possible Causes
NEXT failure	
	Near end connector termination problem
	Short cable with far end connection problem
	Split pair
	External noise source (such as EMI)
	Not Category 3 or 5 rated components
Attenuation failure	
	Excessive length
	High temperatures in components during test
	Connection termination problems
	Link component performance problem
	Not Category 3 or 5 rated components
Wire mapping failure	
	Transposed pairs (may be necessary/deliberate)
	Split pairs (and NEXT will be high)
	Transposed pairs
	Tip/Ring reversal
	Open
	Shorts
Length failure	
	Incorrect setting for nominal velocity of propagation (NVP)
	Actual length exceeds specification
	Opens
	Shorts
	Mixed material
	Not Category 3- or 5-rated components

Figure 4.39.
Recommendations for test failures.

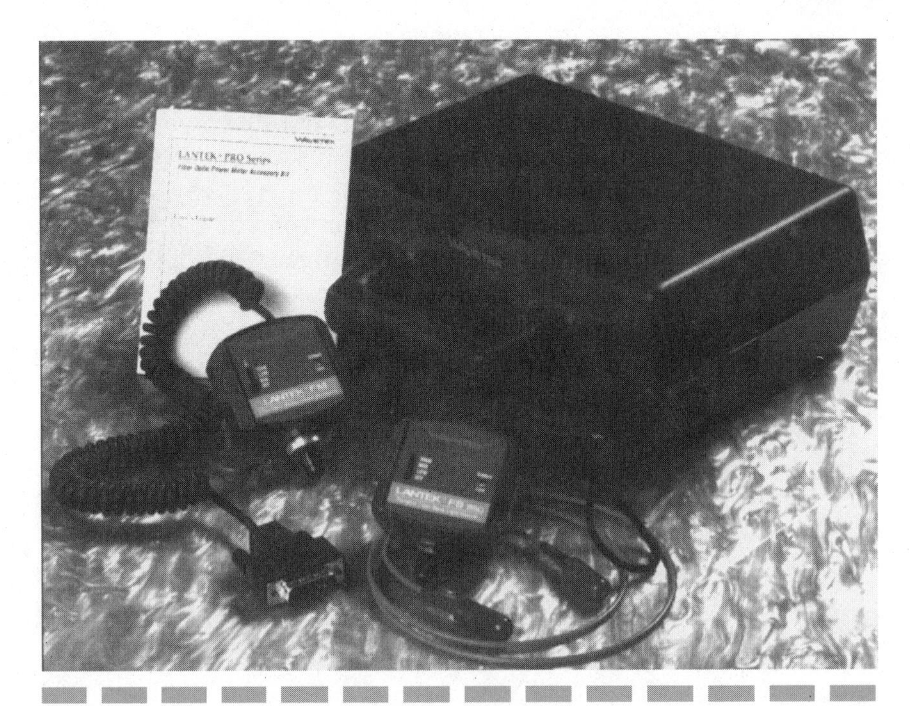

Figure 4.40.
An optical fiber kit for a handheld scanner. (*Wavetek*)

will not repeat signals. It will just view the trunk cable. Because each fiber is single channel and path, it cannot remain connected onto the network all the time. The OTDR will disable the network; it is an important tool in a crisis when a network fails completely. At such times few other tools, especially network software tools which depend on a functioning network, can operate; only noncomputer, nonsoftware tools operate when a network is broken.

The OTDR operates somewhat like radar. It sends a light signal pulse from the attachment point (typically at the connector) down the network media, as illustrated in Figure 4.41. Be sure to notice, however, that this pulse is not compatible with most network protocols. In any event, it will disrupt normal network traffic if the hub remains enabled. In other words, testing a live network is apt to spawn continuous beacons. Because the TDR knows nothing about network protocols, for example in the case of FDDI, the TDR signal will override token and data traffic. Any node that is transmitting a valid frame will soon see that the frame was corrupted. As a consideration to network users, before attaching these devices to a network, inform them that network performance might degrade due to the testing. On the other hand, an individual lobe can be tested without network repercussions if the line is bypassed from the hub or concentrator.

In addition to basic continuity tests, the light meter, feature detector, and OTDR will precisely locate an open or break relative to a cable endpoint, bad splices, poor connections, and signal transmission problems. The difference between these tools is that the light meter requires a source light with a known and stable output at the opposite end of the segment being tested. The feature detector is not a general-purpose tool as the light meter is. The OTDR, on the other hand, generates its own light and reads the light scatter to make an assessment of the fiber. Light sources are generally available in 780-, 850-, 1300-, and 1550-nm wavelengths (standards for optical fiber), although the standard for Fast Ethernet over optical fiber is 850 nm. The FDDI standard is 1330 nm. Some OTDRs require that a reflective terminating connector be placed at the opposite end of the cable or a second OTDR. While testing from both ends might be more precise, it is less convenient because it requires a second person at the other end, or potentially at least two 4-km (or 72-km) trips to set up the other end. By the way, this generally breaks the network during the test period, unless you switch the primary ring to the second channel while you test the primary channel (and hope that the ring does not wrap for any reason).

Furthermore, the OTDR, when used with a camera or a scanner with a serial port attachment to a PC, can blueprint a network. This aids in debugging the all-too-common network failure. For optical networks,

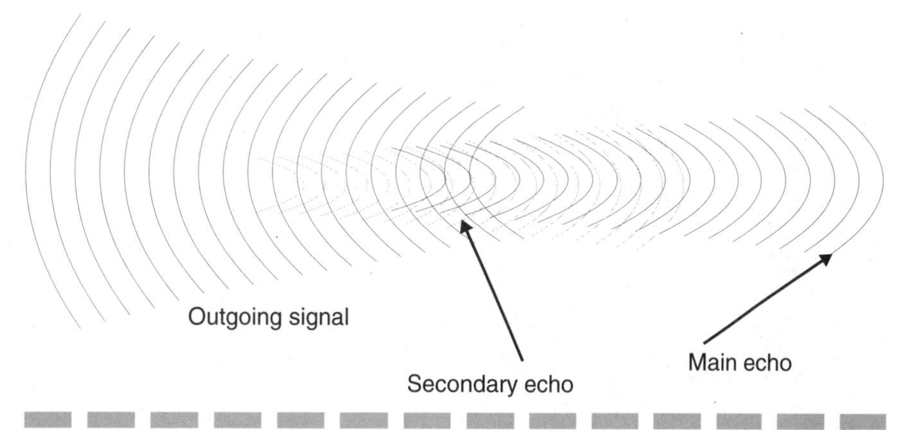

Figure 4.41.
The OTDR signal pulse display.

only a handful of things go wrong. The primary and secondary fibers can be reversed, one of the fibers can be broken, or the signal can be weakened by scatter, optical degradation, or poor splices and connections. Neither the light meter or the OTDR will show that the cables are reversed. That is something best solved with good documentation and use of keyed, shielded dual connectors. Because Fast Ethernet typically uses ST connectors instead of keyed MIC connectors, label, label, label.

If the signal degrades more than the acceptable 3-dB loss, something is wrong somewhere in the length of the fiber. While the light meter and the OTDR both will indicate this problem, the light can help you find it unless you can track it to a separable segment of the cable. If the cable goes bad due to a physical failure, the light meter can only confirm that a problem exists.

The OTDR, on the other hand, can tell you exactly where the flaw occurs. You can follow these length readings to a bad connector, bad splice, or failure of the fiber itself. Connectors can be replaced, splices redone, and cable failures excised from the fiber and respliced. Resplicing is subject to signal loss; the newly spliced cable must remain within specification. Note that when dual fiber is spliced to remove a bad section in one fiber, you should check both fibers for conformity to specification after completing the splice. You do not want to introduce a new problem on another line.

Either of these tools (light meter and OTDR) are effective for qualifying breakout bundles of optical fiber. It is not difficult to tell if a fiber will work or fail. Either it works or it fails for the length and application; both tools can tell you that. Typically, though, a failed fiber in a large bundle is not repaired or spliced, but simply ignored. It is too risky to risk damaging the other fibers. If too many fibers fail, the bundle is replaced in full or a single new line or new bundle is added to pick the load.

The OTDR unit connects onto the network in place of a station. It rarely checks the duplex fibers (RX/TX needed for Fast Ethernet) at the same time. Thus, you might need special connectors or adapters to connect the device to the ring. Also, these jumpers and connectors tend to fail often from wear and abuse.

Signal Characteristics

The pair scanner, light source, light meter, and OTDR must support and be set for the signal "color" or signal propagation velocity that the NICs expect. If you qualify the network or the cable with the wrong bandwidth, it will work for that bandwidth but possibly not for the bandwidth in question. Although multimode fiber will usually support the full range of possible frequencies, single-mode fiber is usually manufactured with optimal signal clarity at the specified signal propagation frequency. One other caveat is that, when testing from opposing ends of a segment, make certain both the light source and light meter are set for the same frequency, or that both OTDRs are expecting the same frequency. You could have the right tools, but use them improperly.

Interpreting Results

All obstructions reflect the signal with various signatures, which are read from a light meter, oscilloscope, or numerical display. These signatures must be learned; their interpretation is an art. Patch panels, splices, connectors, different grades and ages of cable, and (of course) breaks and shorts in the wire are visible. The same holds true for optical networks. Connectors, media attachment units, media breaks or shorts, excessive bends, and segments that are too long are visible in the display. On a digital TDR unit (typically the handheld scanner) lacking a visual display, these anomalies are visible as deviations from the normal numerical display or by a red indicator light.

See Figures 4.42 and 4.43 for samples of OTDR results. The reticle grid (vertical and horizontal lines that run through the middle of the oscilloscope display) is set to discriminate at 200 m. Each horizontal grid line is about a meter. The large spikes represent the test lead.

Figure 4.44 shows a break in a cable. Breaks caused by overextending the bending radius of twisted-pair wire or cuts through a wire pair, or radio-frequency interference from powerful sources (such as transformers, high-intensity lamps, and other electrical cables), show as large bends, spikes, and kinks in the display. Faulty NICs or hubs show up as a waveform different from that of a functioning unit (it usually lacks a large spike. If nonstandard or old media is inserted to replace a defective or damaged section of twisted-pair, the TDR will show this with a large pulse deflection. The test pulse wave will reflect with a dramatic difference in height, and with sufficient practice one can interpret these deviations. The handheld scanners will indicate these anomalies with indicator lights or by reports downloaded to a PC. The oscilloscope CRT also shows distance to the anomaly which is overlaid in the photographic samples. Figure 4.45 shows this common installation problem.

As another example, if the display indicates the network is longer than the specified length, the genderless connectors might be stuck

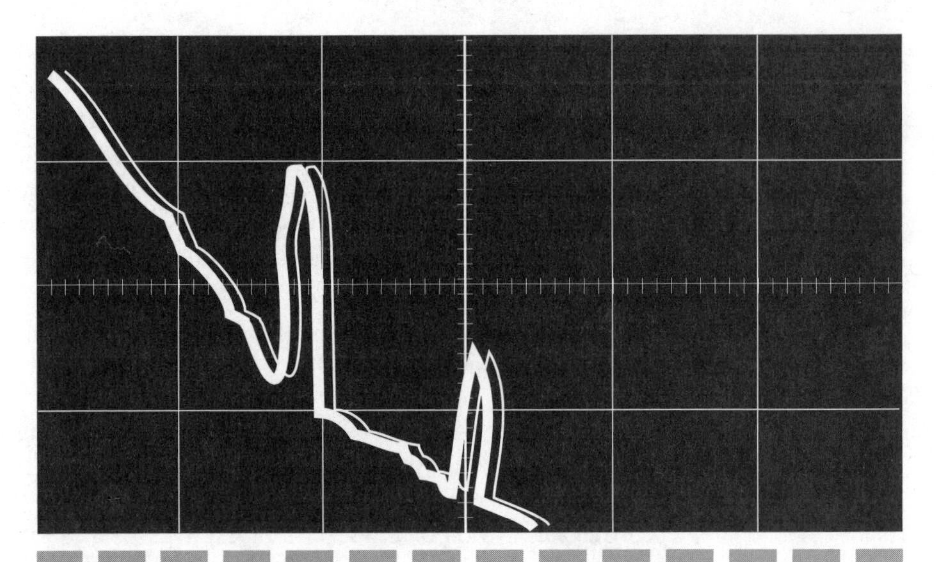

Figure 4.42.
The fiber passed.

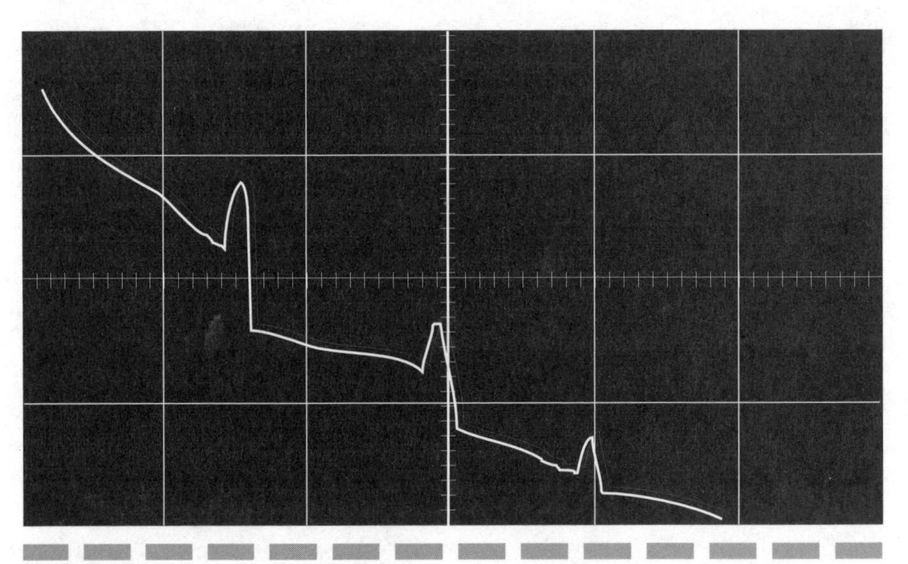

Figure 4.43.
The waveform resulting from the insertion of nonstandard media.

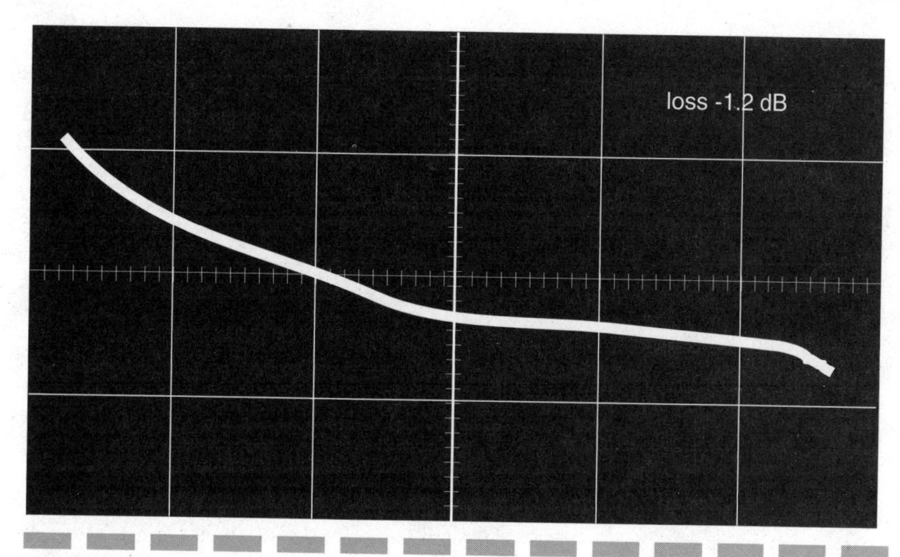

Figure 4.44.
A waveform indicating a break in the fiber (due to lack of features).

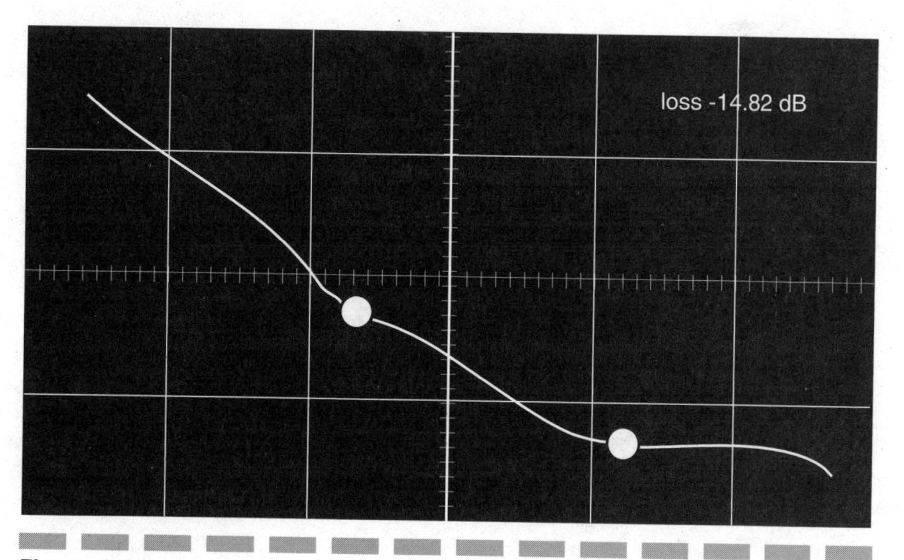

Figure 4.45.
This signature indicates splice loss.

open. The open ring will fail the length test. A second indication of missing ring closure or improperly installed connectors is a display that is mirrored beyond the blueprinted segment length, as Figure 4.46 indicates. This happens because the wave strikes the end of the wire, finds a medium with a different transmission quality, and reflects back down the cable inversely as shown. This is similar to the transmission characteristics of light when it passes from one medium (like air) through another medium (like glass) and back into air. Depending on the reflective angle of incidence, the image will be deflected. A mirror image occurs at the 90⁻ deflection of an improperly terminated end; the image will cease at hubs or stations. Handheld scanners usually will indicate a lobe flaw on a loopback test. The loopback test requires that a special genderless connector is attached at the opposite end of the lobe wire to return the signal originating from the scanner back to the scanner.

Without an OTDR, the course of action would be to segment the ring progressively until the problem is localized sufficiently. Because Fast Ethernet or FDDI is wired as a star, each lobe or trunk is accessible and can be individually tested. Power-off workstations and they will be logically removed from the ring. Disconnect lobe cables or remove jumpers at the patch panel to physically remove workstations from the ring.

Visual inspections are time-consuming and often fruitless. Cable substitutions often mean pulling another cable, a time-consuming task, and

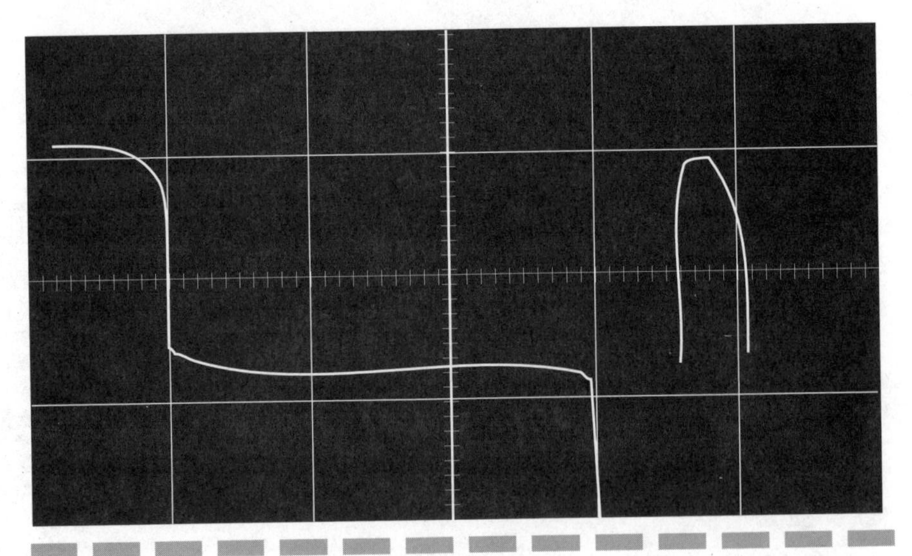

Figure 4.46.
The signature indicates an unconnected cable, or a cable missing a connector.

this is really a waste of time. However, if you pull spares for this eventuality, you can always switch the network over to another cable. If it works, look for dirt on and inside the connector, or scratches on the fiber end. Try the old cable again or mark it for formal testing. The time requirement for such a search is excessive and demonstrates why the scanner is the tool of choice.

Blueprinting

The optical time domain reflectometer and handheld scanner with serial downlink capability are the best instruments to effect physical network blueprinting. This is called the mapping function. The OTDR, in conjunction with a film recorder or screen camera, can illustrate the internal workings of the network wiring or optical fiber. These photographs or data downlinks (from scanner to PC) form a record against which future problems can be compared. It is a map of some permanence. Figure 4.47 demonstrates the type of information that could be logically included in a careful ring-scanner analysis.

Specifically, when the TDR function is combined with protocol analysis, building true topological (although linear) maps representing the network is simple. Reflectometry provides distances to wiring anomalies for each node lobe that can be matched against sent and return protocol frames. The frames provide node names (OSI or Internet addresses) that are matched to physical locations for network nodes. Some anomalies will be connectors, minor flaws, or true faults. Not only is the combination of information provided by these two tools useful for intelligently blueprinting the network, but also it is useful for indicating potential failure points as well.

In actual practice, a network manager and an assistant might use two OTDRs to analyze the network and move them around from location to location. Bad frames, duplicated addresses, and loss of tokens or frames require a protocol analyzer. For more precise verification of network soft errors, the protocol analyzer is the tool of choice.

Protocol Analysis

Few tools provide much insight into malfunctions on multiple segment networks or internetworks. Many network failures can be traced

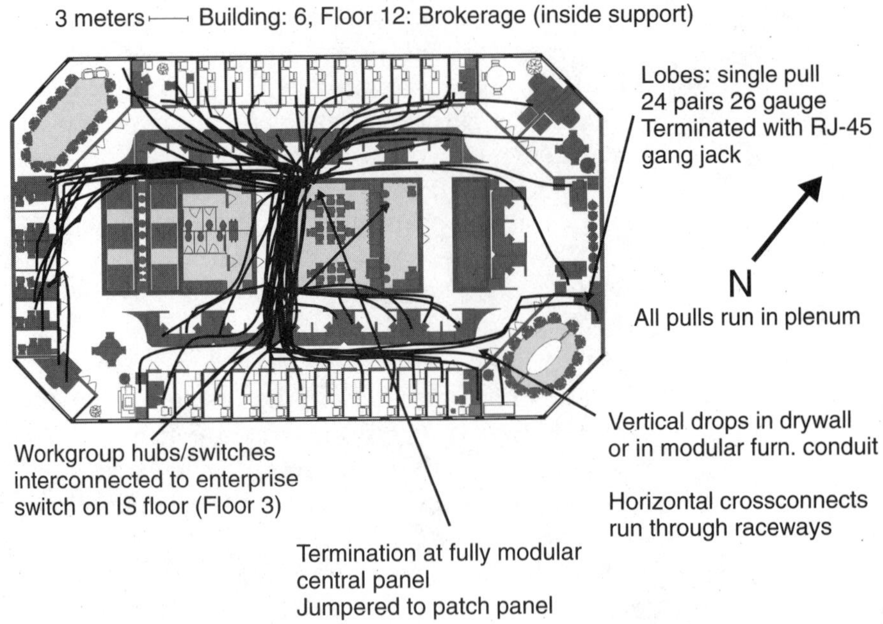

3 meters Building: 6, Floor 12: Brokerage (inside support)

Lobes: single pull
24 pairs 26 gauge
Terminated with RJ-45
gang jack

N

All pulls run in plenum

Vertical drops in drywall
or in modular furn. conduit

Horizontal crossconnects
run through raceways

Workgroup hubs/switches
interconnected to enterprise
switch on IS floor (Floor 3)

Termination at fully modular
central panel
Jumpered to patch panel

Figure 4.47.
An annotated blueprint.

to software that transcends the physical layers of Ethernet, network loading characteristics, or firmware (which is software loaded into a specialized hardware processor). Comprehensive testing requires protocol analysis, which is the process of capturing packets from the network and decoding them so that a person can read and make sense of them. Protocol analysis also includes calculating traffic load and counting packets to generate performance statistics.

Protocol analysis is based on the collection of relevant network operating statistics like throughput rates, collision rates, source and destination addresses, and overhead. Protocol analysis is a process where you view the transmission protocols (most likely standard 64- to 1518-byte IEEE 802.3 frames in the case of Ethernet TCP/IP data-layer components such as MACs and datagrams), and increasingly, client/server messages as the fundamental transmission unit. However, in multiprotocol environments supporting multiple NOSs, you need to decode protocols such IPX/SPX, RIP, SAP, FDDI, Frame Relay, Token-Ring, or a mix of protocols tunneled inside Ethernet packets.

Although you might think that the protocol analyzer is the only tool for protocol analysis, there are a number of options now available. There are built-in partial solutions in UNIX, such as perfmon, netstat, and trace. More functional software solutions include a number of software-only protocol analysis tools, such as LAN Pharoah and LANalyzer for Windows. These tools nonetheless require an NIC that supports the software and the transmission speed and provides promiscuous data capture. Many common NICs do support these features. Other protocol-analysis tools consist of network communication protocol-based tools such as Simple Network Management Protocol (SNMP) and Remote Monitoring (RMON). These services are typically built into the NOS software or hub-based hardware. Additionally, network management stations (NMS), such as NetView, OpenView, Hermes (Microsoft SMS), or Spectrum, provide an integrated UNIX or NT platform for protocol analysis and network management control.

There is one serious caveat for builders of high-speed Ethernet networks. At this time there are few protocol analyzers qualified for 100Base-T, 100Base-FX, 100Base-TX4, or 100BaseVG-AnyLAN. Few can test switched LAN segments even at 10 Mbps, let alone 100 Mbps. While most vendors promise that their 100-Mbps hubs are SNMP-, SNMP-II-, RMON-, and perhaps even RMON-II-compliant, you will want to test these claims carefully to ascertain that they are manageable with your network management stations. Many vendors (such as Cabletron, Novell, HP, and IBM among others) are sharing information, network agents, and network management station software to test interoperability for the end users. Compatibility is extremely important to managers of large networks. Eventually, there will be 100Base-T4-compliant analyzers; there might even be tools for 100BaseVG-AnyLAN. Currently, your best bet is to see that the nodes and hubs will integrate into your network-management environment through SNMP or RMON.

The process of locating intermittent shorts or failures with hardware testers is a random proposition. Such failures require more analytical techniques, including protocol analysis. It is the right tool for locating intermittent shorts or failures, and it provides information to solve network problems through the collection of network traffic data, recording the status of many different variables and converting them into statistics. Most of the newer tools have a sophisticated statistical package that can chart numerical results for visual analysis. Protocol analysis is very important to verify bridge, router, and gateway functionality, particularly now that they are so complex.

SNMP and RMON Protocols

A centralized network management station is the master repository for all managed agents and the management information (data) base (MIB). It controls network agents via commands issued and transported at the datagram level. No logical connections are maintained between the NMS and any agents, thus lowering network overhead. SNMP can trigger alarms based on captured agent information in the SNMP MIB and unsolicited alert information from a network agent. However, the implementation of SNMP does not detract from the more rigorous goal of a comprehensive network management tool to provide management information on complex networks. Ordinarily, SNMP does not provide information across intermediate nodes. To meet this limitation, remote monitor-based tools collect and transport network performance data over various communications channels to bypass bridge or gateway limitations. The table in Figure 4.48 shows the functional support and overlap in the different SNMP and the RMON MIBs.

The growth in internetworks and more complex Ethernet LANs typically requires protocol analysis across the intermediate nodes. The remote (*network*) *monitoring management information* (*data*) *base* (RMON MIB) defines the next stage for network monitoring with more internetwork fault diagnosis, planning, and performance tuning features than any current monitoring solution. RMON uses SNMP and its standard MIB design to provide multivendor interoperability between monitoring products and management stations, allowing users to mix and match network monitors and management stations from different vendors. The RMON MIB enhances current remote monitoring agents with:

- Additional packet-error counters
- Historical trend graphing and statistical analysis
- Traffic matrices
- Thresholds and alarms
- Filters to capture and analyze individual packets

RMON MIB agents work on a variety of intermediate devices, such as bridges, routers, switches, hubs, and dedicated or nondedicated hosts. It is available on platforms specifically designed as intermediate network management stations. An organization might employ many devices with RMON MIB agents (one or more per network segment or wide area link) to manage its enterprise network. Each solution might offer differ-

Service	RMON	MIB II	Hub	Bridge	Host
Interface statistics	•				
IP, TCP, UDP statistics		•			
SNMP statistics		•			
Host job counts					•
Host file system information					•
Link testing			•	•	
Network traffic statistics	•		•	•	
Host table of all addresses	•		•		
Host statistics	•		•		
Historical statistics	•				
Spanning tree performance				•	
Wide area link performance				•	
Thresholds for any variable	•				
Configurable statistics	•				
Traffic matrix for all nodes	•				
Host Top N studies	•				
Packet/protocol analysis	•				
Distributed logging	•				

Figure 4.48.
The focus of various common MIBs.

ent portions of the RMON MIB. Figure 4.49 shows a stand-alone RMON hardware solution.

Many devices, especially older or non-IP products, do not include network management agents and cannot communicate with SNMP consoles except through echo packets. RMON MIB agents monitor every network device and are often the only way to extend network management to such otherwise unmanageable devices. Although many SNMP management stations periodically send echo packets to check the status of each device, this large number of echo packets does increase base-level traffic and potentially could adversely affect performance of the network and the SNMP console. The proxy management capability of an RMON MIB remote monitor is useful when the remote network is connected

Figure 4.49.
A typical RMON hardware agent (TEC).

over a wide area link where periodic polls of devices would consume an unacceptably large percentage of bandwidth. The IETF has planned for future growth in the RMON MIBs by designing its format for easy extendibility to other types of local area networks (such as Token Ring and FDDI) and wide area networks. Because an organization might have several management stations and there can be multiple people managing parts of large network, the remote monitoring agent must work with several management stations concurrently. Using the RMON MIB, each management station identifies the resources it is using in the agent so that multiple tasks are completed in a timely manner.

The MIBs are not simple databases easily presented in a book. The specifications for MIBs typically require 40 or more pages of pseudocode in an IETF RFC. Furthermore, a user of a network management station cannot get to a particular value in an MIB unless it has been coded into the NMS software. In fact, MIBs are complex trees of objects where each single object might be represented by multiple tables. However, each row in a table of MIB values must include an index or a value that uniquely identifies the row so that it is addressable. The RMON MIB specifies that row indexes in many tables start with 1 and proceed sequentially. Because the management station can usually predict the range of desired indexes, a single SNMP command can fetch multiple data requests for better network bandwidth efficiency.

The RMON MIB was developed by a working group of the IETF. The working group included Carnegie-Mellon University staff and people from vendors of network management software and hardware. The RMON MIB is organized into nine optional groups. Compliance with the RMON MIB standard requires only support for every object within a selected group (RFC 1213). Because each group is optional, a number of RMON MIB agents might be used for different purposes; one can be used to trace network structure, another for tracking alerts, and one for maintaining a complete host table. Other uses are possible. Figure 4.50 shows the tree structure of the RMON MIB and the nine leaves that refer to the nine RMON groups.

The statistics group provides segment-level network statistics. These statistics show packets, bytes, broadcasts, multicasts, and Ethernet collisions or ring alerts on the local segment, as well as the number of occurrences of dropped packets. Each statistic is maintained in a 32-bit cumulative counter. The number of Ethernet collisions that are detected by the agent depends on whether the MAU is receiver-based, as it should be for accurate data gathering. Note that Fast Ethernet networks can easily carry more traffic during the specified intervals than can be saved to disk or analyzed in real time.

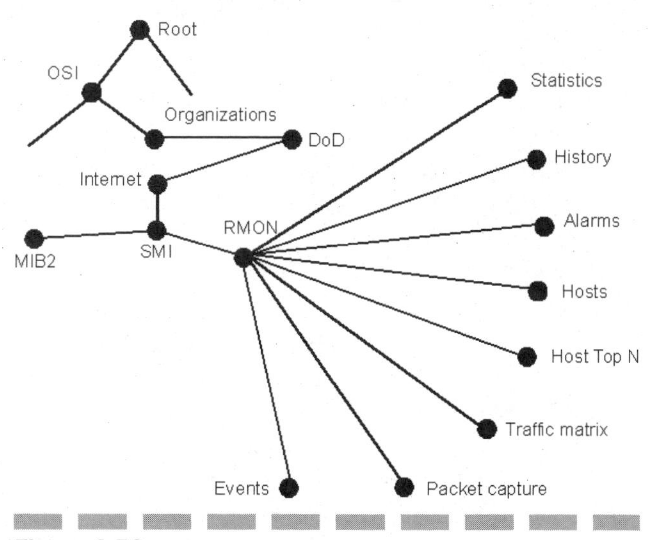

Figure 4.50.
The tree structure of RMON MIB.

The RMON MIB has error counters for five different types of packets: undersized packets, fragments, CRC/alignment errors, jabbers, and oversized packets. These counters provide useful network management information not provided by typical interface cards. For example, industry-standard cards will provide only two separate counts of CRC and alignment, and will not count well-formed packets that are either too small or too large. These undersized and oversized packets are counted by the RMON MIB agent because they usually indicate configuration problems in the transmitting station. Such packets will usually not be passed from the receiving NIC software driver; this results in failed transmissions that are otherwise not indicated.

The history group provides views of the statistics maintained in the statistics group with the exception of packet size distribution, which is provided only on a real-time basis. The history group features user-defined sample intervals and bucket counters for customization of trend analysis. The RMON MIB specifications recommend two trend analyses. The first recommendation is for 50 buckets (or samples) of 30-second sample intervals, for a total time interval of 25 minutes. The second recommendation is for 50 buckets of 30-minute intervals, for a total time of 25 hours. Users can modify either of these or add additional historical time-interval studies; the total number of such statistics is limited by the resources available to a specific agent. The sample interval can range from one second to one hour, creating the opportunity for long historical studies.

A host table is a standard feature of most monitoring devices. The RMON MIB specifies a host table that includes node traffic statistics, including packets sent and received and bytes sent and received, as well as broadcasts, multicasts, and error packets sent. In the host table, the classification "errors sent" is the combination of undersized or oversized packets, fragments, CRC/alignment errors, and jabbers by each node. In addition, the RMON MIB includes a host timetable that shows the relative order in which each host was discovered by the agent. This index entry improves performance and reduces network traffic.

The Top N group extends the host table by providing sorted host statistics, such as the top 20 nodes sending packets or an ordered list of all nodes according to the errors they sent over the last 24 hours. This extensive processing is performed remotely at the agent based on user-defined parameters. This minimizes traffic load on the network.

The RMON MIB includes a traffic matrix at the 802.2 MAC layer. A traffic matrix shows the amount of traffic and number of errors between pairs of nodes, with one source and one destination address per pair. For each pair, the RMON MIB maintains counters for the number of packets, number of bytes, and number of error packets between the nodes. Users can sort this data either by source or destination address.

The alarms group provides a mechanism for setting thresholds and sampling intervals to generate events on any counter or integer maintained by the agent, such as segment statistics, node traffic statistics defined in the host table, or any user-defined packet match counter defined in the filters group. Both rising and falling thresholds can be set, and in fact both can indicate network faults. For example, crossing a high threshold can indicate network performance problems; crossing a low threshold might point out the failure of a network backup scheduled to occur at midnight.

Thresholds can be established on both the absolute value of a statistic or a "delta" value, so that the network team is notified of rapid spikes or drops in a monitored value. This is also true for SNMP as well as RMON. The filters group features a generic filter engine that activates all packet-capture functions and events. All the other groups depend on the filters group, and it is critical to many of the advanced functions of an RMON MIB agent. The filter engine fills the packet capture buffer with packets that match filters activated by the user. Any individual packet match filter can activate a start or stop trace trigger. The contents of the trace buffer are controlled by any combination of user-selected filters. Sophisticated filtering provides distributed protocol analysis to supplement the use of protocol analyzers. The monitor also maintains counters for each packet match for statistical analysis. Alarms can trigger an event to the log using an SNMP trap. Although these counters are not available to the history group for trending, a management station might create trends from these counters through regular polling of the monitor.

The packet capture group depends on the filter group. The packet capture group allows users to create a number of capture buffers and to control whether the trace buffers will wrap when full or stop capturing. Buffers can be allocated in various ways by user parameters. The RMON MIB includes configurable capture slice sizes to store either the first few bytes of a packet where the various protocol headers are located or, at the limit, to store the entire packet. The default slice setting is the first 100 bytes.

The events group provides users with the ability to create entries in the monitor log and/or SNMP traps from the agent on any event of the user's choice. Events can originate from a crossed threshold on any integer or counter, or from any packet match count. Vendors often add features through the network management station software or through proprietary extensions to the MIBs. The log includes the time of day for each event and a description of the event. The log wraps when memory is full, so events might be lost if they are not uploaded to the management station periodically. The rate at which the log fills depends on the resources the monitor dedicates to the log and the number of notifications the user has sent to the log. It is also a function of how much memory is allocated to network management overhead, and the number of user-configured thresholds, traps, and alarms.

Traps can be delivered by the agent to a multiple number of management stations that each match the single community name destination specified for the trap. An RMON MIB agent will support each of the five traps required by SNMP (RFC 1157): link up, link down, warm start, cold start, and authentication failure. Three additional traps are specified in the RMON MIB: rising threshold, falling threshold, and packet match. By the way, it is a waste of resources to set thresholds, traps, and alarms already hard-coded into the basic management features. Make sure you know what features are already coded so that you do not create double (or more) work for not only the remote agents, but also the network management station.

However, most LAN monitors do not have the capability to provide WAN cell and stream decodes, although some vendors sell multiprotocol portable decoders for LAN and WAN traffic. SNMP is useful because it supports RMON, alerts, alarms, and thresholds. The inherent limitation of any in-band management is that bottlenecks can prevent timely delivery of information.

RMON is useful because it uses remote probes beyond the view of most protocol analyzers. The agents often retain traffic and performance data on the device until specifically polled, rather than broadcasting it across the network; this is for performance reasons. RMON devices can be polled to extract historical information. There are nine groups of MIB objects:

- Statistics: This object measures probe-collected statistics such as the number and sizes of packets, broadcasts, and collisions.
- History: This object records periodic statistical samples over time that can be used for trend analysis.
- Alarms: This object compares statistical samples with preset thresholds, generating alarms when a particular threshold is crossed.
- Host: This object maintains statistics of the hosts on the network, including the media access control addresses of the active hosts.

- HostTopN: This object provides reports that are sorted by host table statistics, indicating which hosts are at the top of the list for a particular statistic. Note that host does not mean a mainframe, but any RMON server.
- Matrix: This object stores statistics in a traffic matrix, recording information about conversations between host pairs.
- Filter: This object allows packets to be matched according to a filter equation.
- Packet capture: This object allows packets to be captured when they match a particular filter or threshold value.
- Event: This object controls the generation and notification of events, which might also include the use of SNMP trap messages.

RMON comes with a hefty price in terms of CPU requirements, RAM, disk space for storing statistics, and substantial buffer space to actually capture packets and packet information in real time. Note that RMON (like SNMP) does skew the performance of what you are trying to measure, because the measurement process requires substantial resources.

Network Management Stations

Network management station (NMS) is the official term for open systems tools that run on a workstation or PC and collect SNMP, SNMP-II, RMON, and RMON-II statistics. Many hubs, such as the Synoptics 3000 and 5000 series, include a software module for configuration and management. My use with this NMS has been limited to configuration of the channel utilization and enablement of the hub. I collected some basic performance statistics for installation quality assurance and performance loading through a laptop running Windows. As there were no serial terminals available—and the Bay Networks hubs require one—someone else suggested using the Windows Terminal applet as the only means to emulate the missing terminal hardware. The laptop was connected to the 9-pin serial port on the Bay Networks (Synoptics) hub with a LapLink hydra cable, and eventually I found the right combination of serial port assignments, correct combination of plugs on the LapLink cable, the speed settings that worked, and a parity assignment that worked reliability.

Cabletron, HP, and IBM also provide hub management modules that support different combinations of SNMP and RMON, which is the way to go with Fast Ethernet and switched LANs. These companies also supply software that runs on a UNIX workstation for network management. They provide Spectrum, OpenView, and NetView, and there are many other companies competing to simplify the role of network management with better tools.

The Stand-Alone Protocol Analyzer

The protocol analyzer is a computer workstation that is a network node. It requires transceiver access through a single tap of a multiport unit. The analyzer can send and receive Ethernet packets just like any other workstation. The analyzer watches the network for all Ethernet packets, not just for signals directed to its node. The protocol analyzer is said to be promiscuous because it eavesdrops on all nodes on the network. It is a complicated tool.

Commensurate with this complexity, it provides detailed information to trace complicated problems. A protocol analyzer listens to the signal on the local network and finds a packet with a known source and destination. It must also listen and identify jam, collision, and backoff signals on the local network. Furthermore, this tool must be able to identify packets derived from and destined for remote gateways. Anonymous packets must be filtered and captured for analysis and eventual problem resolution. An Ethernet network carries traffic for many different purposes, such as:

- File services and access
- Print services
- Fax services
- E-mail
- File transfer
- Client/server
- Database access
- Computer telephony
- Backup
- Software installation
- Application code storage

- Process integration
- Virtual terminal support
- Network file system support

The consistency of traffic varies by load rate, by source and destination, and by peak characteristics. Control of the network and proper management require an understanding of the flow of information. Because hardware testers only indicate the status of the physical plant, another tool like the protocol analyzer provides input for network use, load, consistency, resources distribution, and implied resource allocation decisions.

Network Information

Network management software or a stand-alone protocol analyzer dedicated only to the task of watching the network is an important tool for network maintenance and evaluation. Because Fast Ethernet communication is a statistical process (it is a nondeterministic, and hence probabilistic, medium) because a tool is required to measure a medium that "usually works." Problems on Ethernet are statistical and usually appear and disappear, or range from better to worse without anything being "wrong." Also, problems lie dormant until network traffic increases to a statistically critical level before emerging as crises. For this reason, statistical and software tracking is essential. It is also useful if a protocol analyzer can actively page a person when trouble is noted. To date, this feature is not provided by off-the-shelf solutions.

Network Performance Questions

The network administrator must view the network as a highway carrying data traffic in order to resolve higher-level network bottlenecks and Ethernet traffic failures. The network administrator seeks to answer operational questions as listed to evaluate the operational efficiency of any network. Understanding why these questions are relevant is the subject of the next section.

- What do stations transmit on the network?
- Do stations correctly transmit?

- Do stations respond correctly to transmissions?
- Do stations defer to the busy network?
- What is the traffic volume?
- What is the traffic consistency?
- What is the length of messages?
- What is average peak loading?
- What is full capacity?
- What is the average intermessage timing?
- What is the average wait delay?
- Which stations talk to each other?
- Do messages have frame errors?
- Are some stations unable to reach other stations?
- Which nodes cause problems?
- Which section or segments are overloaded?
- Are gateways and repeaters performing to specifications?

If you dealing with an internetwork, the previous questions are pertinent, but you also want to know about end-to-end network latencies (the real killer of performance) and subnet and backbone loading. Obviously, an internetwork is a product of routers, and some integrated management of that router web is desirable. While some organizations have built internetworks exclusively with routers from one company to simplify management and configuration, the internetwork will be changing faster than these vendors can singly support in many cases. As such, you will want or need intermediate nodes—they could be routers or they might be switches—from other vendors. The protocol analysis issue becomes one of uniform data gathering. SNMP and RMON help, but you still need to know the questions, which are:

- What is the overhead for the router protocol?
- How long does it take for a router (or switch) to converge?
- Is any traffic caught in an endless loop?
- How long does it take to reconfigure after a route failure?
- Do the spanning trees work?
- Which routers are overloaded or close to capacity?
- Can subnets and loads be rationalized?
- Are the routers talking to each other?

The Heisenberg Uncertainty Principle

If the monitoring equipment actively requests and captures information, the process itself skews the results. It is quite likely that file service requests, information probes, and process swap requests performed in the pursuit of network status data will itself generate network loading. The essence of the Heisenberg Uncertainty Principle is that the active process of observing an event alters the outcome of that event. Passive monitoring, on the other hand, watches each broadcast packet without accessing or affecting the network. This eliminates performance losses and potentially obviates any speed limitations that the monitoring actually seeks to chronicle. Unlike most vendor-supplied software tools, most protocol analyzers provide passive monitoring.

Vendor-Specific Limitations

Most vendors of network workstations provide resident software tools to track network access and usage for individual workstations. Some vendors even provide software that will measure network usage. Choose software and hardware that conforms to the SNMP standards. This provides an agency address for SNMP data collection. Hardware and software lacking in this regard are effectively invisible to the management protocols. If these software tools are accurate and not approximate, an analyzer might be unnecessary.

However, vendor-supplied network-monitoring software tools will exhibit the same flaws and faults as their workstations and overlook some types of problems. Often Ethernet networks are built of many parts from different vendors over a period of time. As a consequence of this typical growth pattern, few types of node equipment can monitor a mongrel network with any accuracy. Therefore, a specialized device is required to capture and parse all types of network packets. The protocol analyzer fills this function because it is protocol-independent. The analyzer can view all IEEE 802.3 and Ethernet packets that conform to TCP/IP, HPnet, XNS, DECnet, OSI, or any other "standards." The tool passively captures a burst signal containing an Ethernet frame, then parses this packet to whatever the upper- or lower-level protocols dictate. Within this framework, TCP information and IP control information, for example, can be viewed without using questionable network-level software.

Some workstation vendors build analyzers as part of their product line. In general, it is a good management practice not to rely on single solutions in this specific case; it is desirable generally to install networks with components from a single vendor. However, test equipment is sometimes better acquired from a competitor to your network device vendor. These backup and alternative methods are reasonable. Therefore, consider purchasing a protocol analyzer from an independent vendor so that any vendor-specific problems or incompatibilities with Ethernet, Fast Ethernet, and demand priority standards can be uncovered.

Understanding Network Performance

Ethernet is a purely statistical transmission medium. Network node stations transmit when they see a free line, and collisions occur when two or more nodes see the free line simultaneously. Because Ethernet (versions in common usage) assumes a minimum interpacket spacing, there can be no "tailgating." Additionally, traffic speed is enforced at 100 Mbps. Overloads and traffic problems tend to occur at the nodes themselves, barring cable hardware problems. Network problems can be grouped in a hierarchical progression. It is usually worthwhile to check for node hardware problems before searching for network hardware failures. Likewise, it is usually best to search for node software problems before global bottlenecks, incompatibilities, or failures.

In order to monitor network performance, it is important to know how many nodes there are on the network, and which actually transmit. The first order of network problems occurs when nodes do not transmit correctly. This is usually indicative of improperly specified hardware or network software, or defective components. Likewise, nodes might not respond to transmissions (including collision jam signals). The analyzer gathers measurements to pinpoint these problems. Once the deterministic network hardware problems are resolved, the second order is the actual traffic load. Each and every piece of equipment might function correctly. Systems might be reliable. Nevertheless, a network might perform slowly and erratically. This is usually indicative of an overloaded transmission channel.

The most significant network measurement is the packet count, which tells how many packets, good or bad, have been transmitted on the network. This count can be subdivided into good packets and bad packets, by source and destination, by characteristics of packet length,

	Good packets	Bad packets
Length	average	Collisions
Length	variance	Short
Frequency	average	Long
Frequency	variance	Misaligned
Frequency	mode	Frame errors
Source		FCS errors
Destination		Protocol errors
Source-destination matrix		Stray signals
Network overhead		Dropped or lost
Router overhead		Jams, time-outs
RIP/SAP and announcements		Wrong addresses

Figure 4.51.
Network packet measurements. These measurements are also grouped by packet source and destination.

type, frequency, or by error condition. Figure 4.51 groups these items for practical implications.

Basic network monitoring is the process of capturing packets during a known time frame, because problems occur within an interval of time or can be pegged to a certain event in time. Therefore, packet volume (or throughput) is an important measurement. The consistency of volume is another important unit, because inconsistency creates traffic bottlenecks that frequently do not clear; traffic consistency will be discussed in the next chapter. If network problems occur and the assumption is that the network is overloaded, volumes and peak volume statistics might validate this theory. However, it is important to compare observed volume to a theoretical capacity.

Other areas of network transmission problems are the degenerative transmission situations that occur when packets are defective, too long, too short, misaligned, wrongly addressed, or incompatible with the network standard. The search for the cause of these problems transcends statistical measurements, because bad packets have to be retransmitted until they are received correctly. Bad packets can flood a network and halt its effectiveness. In order to expose any sources of bad packets, defective packets must be examined for content and the packet fields should be parsed. In fact, protocol variances usually show in address discrepancies, length or type of field variations, or the LLC and upper-level protocol fields contained within the actual Ethernet data field. If underlying causes for corrupted packets, address discrepancies, or mismatched upper-layer protocols are to be exposed, packets must be dissected.

Collisions

The most meaningful statistic (and the one most often provided by resident network Ethernet software) is the collision rate. Collisions, you recall, occur when two or more nodes overlap their transmissions. A high collision rate is indicative of many problems, including any or several of the conditions listed below:

- Overloaded network
- Too many nodes (exceeding specification)
- Too much traffic
- Network backbone too long
- Hubs cascaded too deep
- Interface errors with switches
- Chattering, jittering, or jabbering transceivers
- Defective transceivers
- Defective Ethernet cards
- Defective software

Corrupted Packets

Comparisons of rates for short, long, and corrupted packets are another statistic of importance. Damaged packets imply a collision, a defective Ethernet card or transceiver, problems with the coaxial cable, or outside noise. A damaged packet usually has a bad checksum or cyclic redundancy check (CRC). Short, long, or damaged packets imply faulty software or hardware at a workstation. Statistics on corrupted packets should focus attention on the root causes of problems rather than on some ephemeral "global" problem. Excessive numbers of short, long, or defective packets could imply that transmission from a node transceiver was interrupted, or that the transmitting workstation did not realize that the network was busy and incorrectly transmitted information without listening for carrier or other transmissions. There is also a certain number of truncated packets that occur each time there is a collision.

A protocol analyzer with a robust software package eases the difficulty in calculating these items, because it will automatically tally packet counts, calculate percentage rates, and graph trends. Graphics display of these items lends more meaning, and can show trends not otherwise noticeable. However, even with a less satisfactory software package, the basic information can be collected and downloaded to a spreadsheet program for the same results.

Packet Counts

At a minimum, some metrics are required to compare network performance over different time frames. Another network status indicator, in addition to the collision rate and the error rate, is a packet count. This number can be compared against a count from another day to see if unusually heavy traffic is the reason for perceived performance problems. For example, if there were 15,000,000 packets today versus 12,000,000 packets yesterday—a 20-percent increase—the magnitude of this difference would explain why performance has suffered, assuming no significant change in the collision level. An even more useful statistic would be a packet transmission rate per hour, called a traffic rate. While packets per day is a useful comparison, packet counts are more useful when presented in more realistic and comparative units. Counts per second, per minute, or per hour constitute what is typically a more useful traffic rate. These counts allow comparisons during the shorter intervals within which most Ethernet problems occur.

Statistics for Network Planning

The number of good packets is irrelevant until correlated with other information. By combining the number of good packets transmitted (rather than the total packet count) with either the lengths of each packet or average length of transmitted packets, a transmission-level count is generated. This is a bps (bits per second) figure. Every manager needs an indication of how much work is actually accomplished for planning growth, managing new tasks, and maintaining a functioning network. Collision levels are not practical for this because they increase nonlinearly; when the traffic is near the breaking point, you can add one new workstation and the collision rate is liable to double, triple, or even break the network. Usage rates aid planning by providing a nearly independent statistic. It is nearly independent because as the network grows and the load increases, response time decreases and less work can be completed within any interval. Consequently, usage rates are apt to be lower per node on a saturated network than on a nonsaturated network. Usage rates are more consistent.

The usage rates also show type and quantity of work by node. They provide the necessary data to make informed network planning decisions. Chapter 5 presents many options to tune network performance. The problem remains to identify what should be tuned.

Network Traffic Composition

A subset of nodes will have higher rates of usage than other nodes as a consequence of the work performed, or as a result of those nodes' need to access network resources. Different users work different hours, perform various duties that load a network differently, and generate different consistencies of Fast Ethernet packets. Also, too many users and nodes clearly can overload a network and create bottlenecks and slow network response. Indicators of node-specific overloads are:

- High packet-density count (high-level load)
- Lengthy transmissions (high-peak demand)
- Minimum (or less) interarrival times (saturation)
- Slow user response (overloads)
- Packet corruption (overloads)

To understand the root cause of uneven traffic problems, it is important to understand the composition of that network traffic. The pie chart in Figure 4.52 summarizes data from captured packets. Of 3300 packets, 500 were mail messages, 300 were follow-on packets for mail messages, and 2400 were applied to other uses. Only 100 packets were identified as collision fragments. The graphs imply that 25 percent of all network traffic in this sample is mail. Because there were follow-on mail packets, 60 percent of all mail messages average longer than a kilobyte. Note that 40 percent of all messages are brief. It is possible that one message is 0.3 megabytes long (requiring all 300 follow-on packets). This skews the statistics and is one hypothesis to investigate.

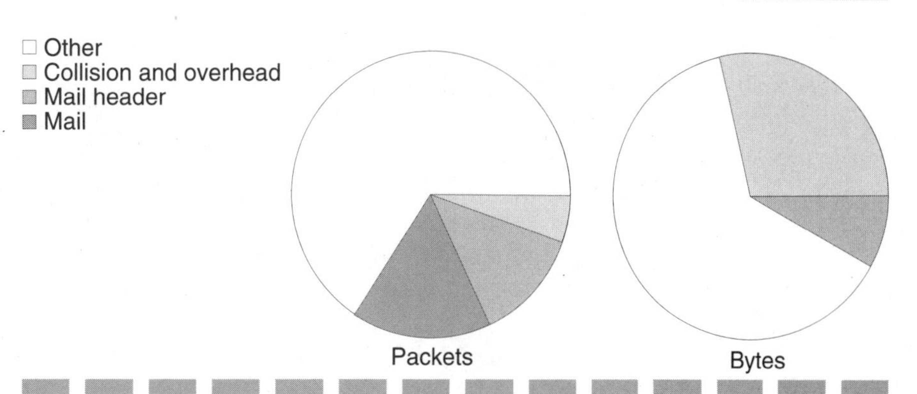

Figure 4.52.
Indicators of network traffic composition problems. The pie charts detail the percentage of network traffic allocated to a particular function. In this example, E-mail creates most of the network utilization.

The analysis does show that most mail messages are fairly brief, and 500 messages are a fairly significant number. If they come from a single node, it might imply that one user is broadcasting globally, and this might be unacceptable. Perhaps this type of transmission should be restricted to certain hours. If most messages are transmitted to a single node, perhaps this transmission level might imply the need for a bulletin board system rather than a mail broadcast. These are courses of action that packet content might suggest. Packet header information could be correlated with each mail message to indicate the actual length, because the header does contain the length of the actual information.

Some processes might demand a significant portion of network traffic. Electronic mail, for example, might be a hidden use of network resources. The implication in this example is that mail is disproportionately overloading the network; while packet counts for mail are not large, the effect of actual traffic is skewed toward servicing mail requests. Likewise, data file uploading and downloading might be a significant load; this can be verified by examining the contents of the long trains of packets sent between specific pairs of nodes. Time-stamping of packets would aid in this type of investigation. Perhaps, like mail, this type of transmission should be restricted to certain hours as well.

A solution, if usage-specific traffic is deemed a problem, might be to identify the need for electronic mail. Another similar source of network traffic is graphic paging, sometimes known as screen refresh. This condition might imply that large blocks of high-density graphic information are being transferred with frequency from one machine to another. Several solutions to these types of problems are possible. First, transfer could be accomplished by tape or floppy disk rather than by the network.

Second, a special link might be constructed to remove the burden from the main network (for example another Ethernet cable or RS-232 line) solely to support this transfer channel. Third, mail could be posted on a passive bulletin board rather than broadcast on the network.

Global and Node-Specific Data Collection

Network software tools are sometimes available on the workstations, network servers, or PCs that are connected to the Ethernet network. Network software tools might count packets and packet collisions, and sometimes identify source and destinations with packet counts. Most often this software provides only single-node information and not global information about all nodes. Both node-specific and global information are required for the best results in network maintenance. This information must be accurately acquired and displayed in useful formats. The minimum functionality for monitoring transmission-level operations required to answer the network manager's questions includes the following capabilities:

- Count packets:
  - all packets
  - short packets
  - long packets
  - corrupted packets
  - good packets
- Display length of each packet
- Display average size of packet
- Count collisions
- Store packets to memory (and/or disk)
- Monitor traffic levels
- Monitor individual nodes, looking for:
  - high levels of traffic
  - high rates of collision

- high rates of retransmission
- type of traffic
- corrupted packets

A global collision rate indicates a global problem such as overloading or network configuration violations. Although a global rate might indicate individual component failure, it is more effective for diagnostic analysis to compare global rates against individual rates at the node level. Node-level collision rates should be uniform unless there is a node-specific problem. Collision rates above the average indicate nodes that are prime causes of collisions. Collisions might also occur specifically among a subset of nodes. Collision statistics for specific nodes allow the network manager to focus attention on root causes of collisions. For example, one subset of nodes might have transceivers or Ethernet cards from the same manufacturer. Problems developed (or, more likely, have not appeared until the network became critically saturated) with this hardware. Or, to take another example, collisions might only occur between two different types of workstations. In such cases, it's likely that the workstations are having difficulty intercommunicating. Problems at this level are common, but are uncommonly difficult to diagnose without node-specific correlation.

Unless single-node statistics can be captured inexpensively and analyzed globally, network software tools are only transient indicators of network problems. Single-node statistics might pinpoint a problem with a particular node without indicating a specific node as the source of problems. Global counts might indicate a local or global problem, but rarely will pinpoint the cause. Unless global information with node-specific corroboration is available, problems cannot be isolated to a specific node. Even a problem as common as chattering transceivers cannot be resolved without such corroboration. Without the specific node information correlated to the global information, excessive transmissions of meaningless packets cannot be traced to their hardware sources except by trial and error or a binary search with physical test equipment.

The network statistics in Figure 4.53 show a high rate of collisions. This global statistic would lead one to believe a problem exists everywhere. Node analysis from simple node-specific software would suggest that both nodes 1 and 3 are faulty, whereas in fact only a single node might be causing problems; both a 20- and a 25-percent collision rate are significant. One node might be demonstrating that it is having difficulty communicating with the other node. More specific information corroborating that all collisions at node 3 were caused by transmission to node 1 would confirm that only node 1 is a problem. The discrepancy of 400 packets could be associated with the runt or oversized frames that would occur if node 1 had a faulty transceiver.

This is actually a very simplistic presentation representative of only small LANs. The matrix gets very large with LANs having more nodes than that, and is downright difficult to view for internetworks. Most protocol analyzers (including LANalyzer for Windows) categorize the top 10 nodes so that you can actually cope with and respond to this information. If you set thresholds and alarms, the information is usually provided to you at the time when you need it, and the presentation is geared to a crisis-mode approach.

However, you do want to baseline normal network performance for two reasons. First, so that you have an idea as to what constitutes normal traffic—packet loads and error rates included—and you also want to know at what level to set thresholds and alarms. If they are set too low, normal peak loads will exceed the thresholds and activate the alarms. If it happens too often, the false alarms will dull your response to true ones. Therefore, you really do need to understand these simple statistics as gathered by most protocol analysis tools. From there you can appreciate the power of the statistical information and the value of alarms.

Going back to the prior traffic example, if usage rates seem localized to certain groups of nodes, or localized to a certain manufacturer's parts or certain types of users, then it might be desirable to localize those groups on subnets or allocate special resources to the groups. Often those groups might not warrant the extra costs associated with the additional resources, or it might be politically unfeasible. It is important to note that Ethernet is a public resource without locks or accounting statistics. The added resources might be applied to improve the global performance incidentally in the attempt to improve performance for a selected group of nodes and users. Capacity is a global constraint, not a controlled resource.

	Node 1	Node 2	Node 3	Network
Output (packets)	12400	15000	15000	42400
Input (packets)	6000	22000	900	37000
Collisions	3000	0	2000	5000

Figure 4.53.
Node-specific data bring global analysis into perspective.

Networks in transition from Ethernet or Token-Ring to Fast Ethernet or 100BaseVG-AnyLAN in particular present potential for bottlenecks at any intermediate node buffering the mixed transmission speeds. It is not enough to look at traffic transmission loads, but also timing delays through the intermediate nodes. Because Ethernet packets typically include a time stamp, and certainly those from a protocol analysis tool do, you want to review delivery times for any performance lags.

General-Purpose Network Flow Information

Figure 4.54 lists some typical questions and corresponding important network statistics by name. The listing, by no means complete, is representative of the most expedient manner to answer each question posed.

Guidelines for protocol analysis are very difficult to construct and not very useful. The process of protocol analysis is one of classifying the ebb and flow of network traffic, and it is a new paradigm. If you notice, the last four questions generated four different answers. This is for the purpose of showing that you can use very different types of information to reach the same answers. None of them are wrong, though some are better than others. The concept that I would like to get across is that you want to create a logical diagnosis of the network situation so that multiple statistics can reinforce and corroborate your troubleshooting and performance analysis.

For example, router buffers might be full because the other LAN is not functioning, or because the traffic load across the router is too high. Many lost and dropped packets do not differentiate a nonfunctioning LAN from an overloaded router, but certainly add to the analysis. A source quench message to a server from a router would decide the issue, but many router protocols do not provide this feature. Instead, a count of packets routed through that router to other networks (but not routed to the suspected LAN) would confirm that any performance problems were with the LAN rather than the router itself. There are no single statistical answers that are so clear and precise as to answer complex network traffic and performance questions. However, when relevant statistics are used in conjunction, you can weave an answer that will be useful.

The next list is by no means a complete listing; it is just representative of that network monitoring information generally provided by a protocol analysis tool, whether a hardware analyzer, a software analysis tool such as LANalyzer for Windows, SNMP or RMON agents, or a network

Typical network questions	Relevant statistics
What nodes transmit?	Source and destination addresses
Do nodes transmit correctly?	Capture error packets with source
Do nodes respond to transmissions?	Simulate with directed packets
Do nodes defer to other signals?	Check collision rate
What is traffic volume?	View utilization rate
Is traffic load consistent?	Increase interval for utilization rate
Are users playing Doom on the net?	Look for Doom signatures
What is normal traffic load?	Baseline key statistics
What is excessive traffic load?	Establish baseline thresholds
Which nodes talk to each other?	Create a source/destination matrix
Is the network segmented correctly?	Review matrix values
Do all stations reach each other?	Any user complaints?
Which nodes are problems?	Look at utilization and error rates
Are any routers overloaded?	Look at router buffer utilization
Are any routers overloaded?	Look for lost/dropped packets
Are any routers overloaded?	Look for source quench messages
Are any routers overloaded?	Count routed vs. nonrouted packets

Figure 4.54.
Typical management questions and corresponding statistics.

management station. The equations for these or the methodology to generate them are explained in detail in the next chapter. This list is included here in order to annotate several of the next charts and detail the applicable uses for the individual counts and rates:

Packet counts	Packet errors	Channel utilization	Capture	Statistics
All packets on networks	All packet/frame errors	Throughput	All packets	Network usage as a percent of capacity
Packets transmitted	Alignment errors	Peak rate	Packets during an interval	Interarrival times
Packets received	CRC errors	Utilization as a percent of capacity	Packets with errors	Latency distribution

Packet counts	Packet errors	Channel utilization	Capture	Statistics
Packets transmitted with collisions	Short packets		Packets with specific errors	Packet size distribution
Packets deferred	Long packets		Packets with certain data types	Node usage as a percent of capacity
Packets per node	Invisible packets		Packets from specific source node(s)	Node usage as a percent of network utilization packet
By source and by destination	Incompatible packets		Packets directed to specific destination	Rate per second (bps or Bps, Mbps or Mbps)
By source to destination (matrix)	Misaddressed packets		Node(s) packets with specific source/destination	Collision rate
Size distribution (-, 64, 128, 256, 512. 1024, 1518, +)			Pairs packets with a certain size	Retransmission rate
Network period peaks node period peaks			Packets without a protocol	Delivery error rate
			Packets with a specific protocol	
			Packets that pass through an intermediate node	
			Routed, switched, and encapsulated packets	
			Packet fragments	
			Packets with SNMP information	
			Packets with RMON information	
			Broadcast packets from routers and gateways (storms)	
			Threshold warnings and alarms	

Network Performance Data

Now that I have identified statistics of interest, a measurement methodology needs to be developed. The worksheet in Figure 4.55 includes sample data for a large, complicated network, and the necessary steps to convert the collected data into applicable information. In this sample, the users experienced a network that is slow and unresponsive; they have complained to the network manager, who must validate or discount the users' claims. As a consequence of a thorough analysis with an analyzer, several unrelated problems are uncovered.

Although node E shows a moderate collision rate on its own packets, it dominates the network collision rate. This "red herring" is immaterial, because the network collision rate is reasonably low at 3 percent. Nodes A, H, K, M, and X have disproportionately high error rates, and should be looked at for cable, hardware, or software flaws. Nodes N and Z probably have bad drivers or old hardware. Nonetheless, nodes B and E might require a subnetting or a revision of the subnet architecture. Instead, if these nodes are servers, you might consider adding another NIC each to these two nodes and relocating some of their client nodes to the new subnets. If you are using stable hubs or chassis hubs, the microsegmentation is easy.

In this example, information has been captured during a 24-hour period on a network with 26 nodes. The column headings represent each node, the number of packets input (i.e. received), the number of errors at the transport layer for these input packets, the number of packets output (that is, transmitted), the number of errors at the transport layer for these output packets, the collision count for output packets, and five columns for rates. The first column shows input errors as a percentage of packets received. The second column shows output errors as a percentage of packets transmitted. The last two columns cannot be calculated until all the other information has been completed, because these rates require network totals.

The following eight graphs compress a large number of network statistics into a clear picture. The network represented is a complex network with one main segment interconnecting four lesser segments; it has 172 nodes in all. The rate of utilization reaches saturation levels so that typical problems are presented to the reader. Saturation is not as likely on switched segments; however, you do need to understand how to read this traffic to create load-balanced segments even with a VLAN architecture.

Figure 4.56 subdivides the network load by source machine type. In this sample, a conclusion to be made is that the Xerox equipment is providing a baseline activity during the course of the day, perhaps consisting of data backup, coprocessing, global modem operations, or mail services. The load on the IBM equipment is counter-cyclic to the normal workday, suggesting some planned off-peak process, whereas the Apple and DEC equipment appear to be servicing users during the normal

Node	Input	Collide	Output	Error	% input	% output	% tot
A	89897	261	37307	5	0.29	0.01	0.18
B	2492034	0	6116816	5	0.00	0.00	29.53
C	192212	251	129838	11	0.13	0.01	0.63
D	561151	0	415580	10	0.00	0.00	2.01
E	15987246	0	10022484169	362252	0.00	48.39	51.85
F	812002	446	747493	25	0.06	0.00	3.61
G	42862	0	44130	2	0.00	0.00	0.21
H	611480	1923	42840	14	0.31	0.03	0.21
I	115363	38	64289	1	0.03	0.00	0.31
J	103	0	84478	867	0.00	1.03	0.41
K	82097	234	40996	14	0.29	0.03	0.20
L	65316	346	10880	2	0.53	0.02	0.05
M	153256	0	120172	94	0.00	0.08	0.58
N	618	452	3140	3	73.14	0.10	0.02
O	1040322	1824	604794	48	0.18	0.01	2.92
P	1334398	347	1315634	9	0.03	0.00	6.35
Q	44748	0	53752	0	0.00	0.00	0.26
R	91380	0	32778	4	0.00	0.01	0.16
S	476121	266	483599	59	0.06	0.01	2.33
T	80594	0	72483	4	0.00	0.01	0.35
U	79780	0	71956	4	0.00	0.01	0.35
V	92126	0	21619	1	0.00	0.00	0.10
W	141428	0	108958	0	0.00	0.00	0.53
X	6018	0	501	3	0.00	0.60	0.00
Y	116901	150	66893	9	0.13	0.01	0.32
Z	33	2	59	7	6.06	11.86	0.00
NET	24709488	6540	20713472	1370			

Figure 4.55.
Statistical worksheet for 100Base-T.

workday. Such assumptions could be verified by packet capture techniques or by learning about the specific processes on the network.

This same information is displayed in a different format by network service usage in Figure 4.56. Most of the peak load is created by user demands during the normal work hours.

The graph in Figure 4.57 presents loading on the main network segment that interconnects with four other segments. This traffic is identical to that represented in Figure 4.56 as the overlay implies, although some of the load is localized on the subnets and therefore does not appear on the main net. The data used to construct the graph in Figure 4.58 are repeated by the histogram in Figure 4.59, which displays traffic by packet size.

Several upper-level protocols and errors are overlaid to indicate the common packet sizes. In this sample, TCP/IP is the upper-layer protocol. Errors are represented by packets that are less than the 64-byte minimum (which represent collision fragments) or those packets that exceed 1518 bytes (which represent transceiver chatter).

The network collision rate correlates to the peak load in the last time-period charts. Because the collision rate is nonlinear, it jumps when traffic builds above 40 percent of network capacity. This loading is obvious during peak business hours, and drops off during lunch and afternoon slack times as Figure 4.60 illustrates. Interarrival times should be a minimum of 96 ms (9.6 ms at 100 Mbps). Interpacket spacing of less than 96 ms at 10 Mbps (9.6 ms at 100 Mbps) represents collisions, signs of hardware

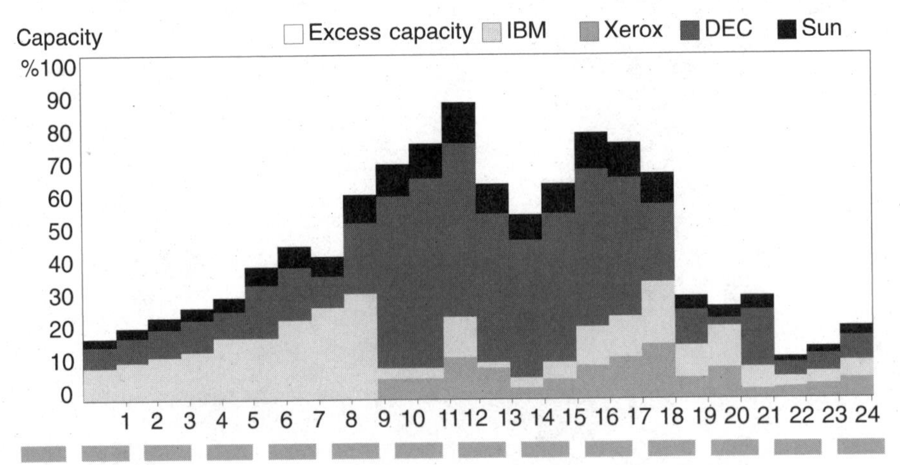

Figure 4.56.
Peak load tabulated by source computer during a 24-hour period.

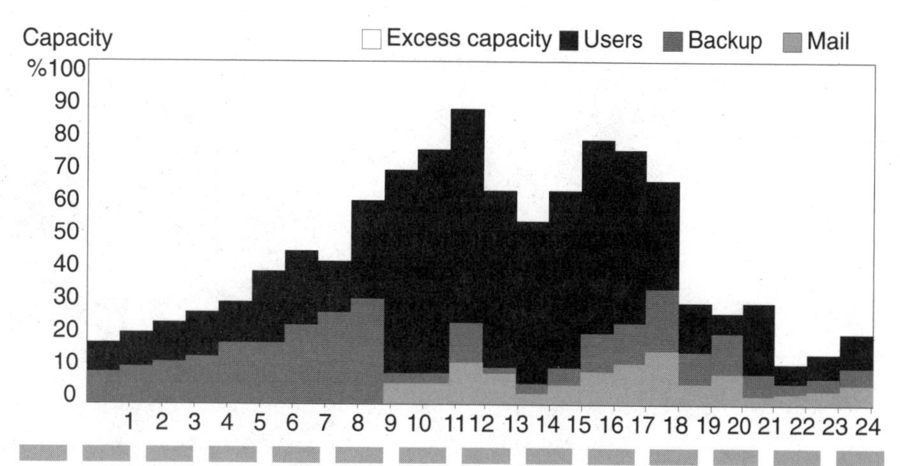

Figure 4.57.
Peak load capacity by usage during a 24-hour period.

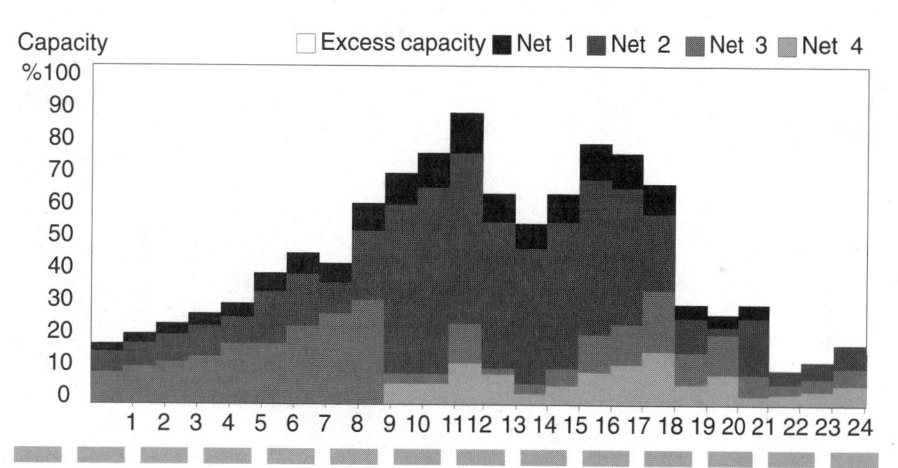

Figure 4.58.
Peak load tabulated by network source during a 24-hour period.

defects, or mismatched network analysis in a multispeed or multiprotocol environment. The network interarrival time in Figure 4.60 implies a 35-percent collision rate, thus duplicating information in Figure 4.59.

Network access latency (as shown in solid black) in Figure 4.61 is the time required for a request to be filled. The information is superimposed on the interarrival time (as shown in gray) for comparison. Latency times provide less important information than collision rates or error rates, although they can confirm whether the problem is overload or

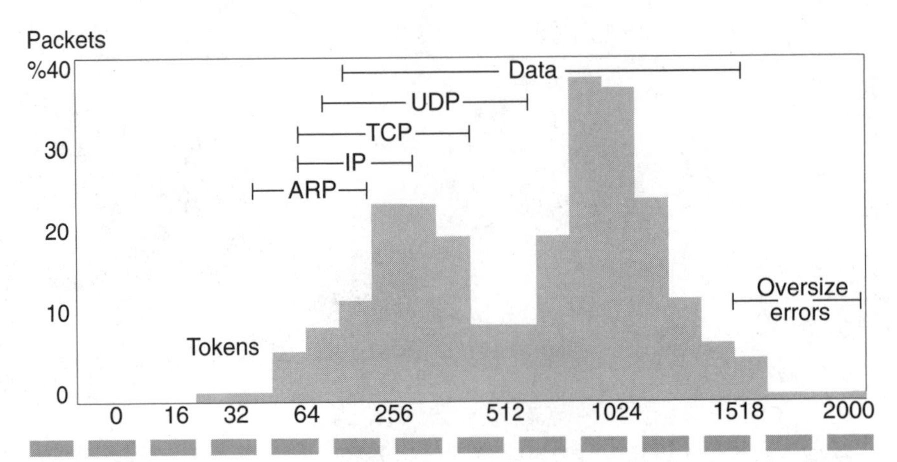

Figure 4.59.
Distribution of packets by size. Each different size represents a packet for a different purpose.

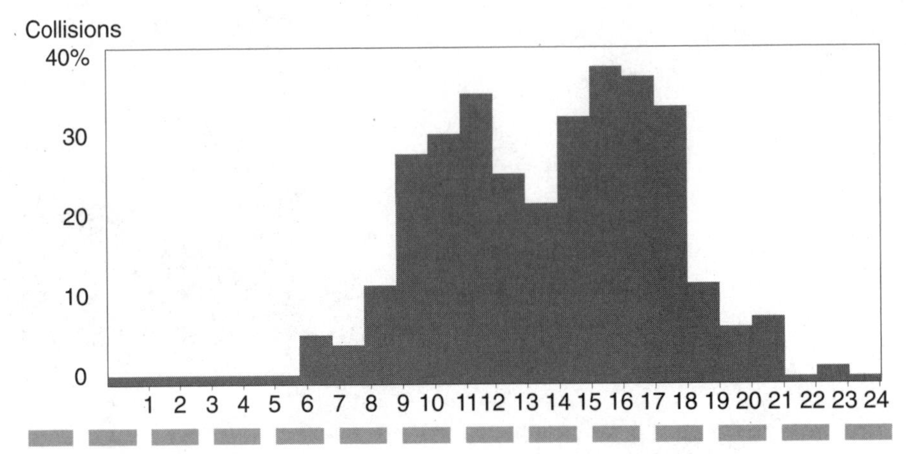

Figure 4.60.
The collision rate during a 24-hour period.

hardware. In this case, because the latency shows a bell curve, the implication is that the network functions correctly except when the traffic load is too high, and latency shifts outward.

Node Simulation

When statistics fail to identify network problems, node simulation on a protocol analyzer is a powerful methodology. Specialized software constructs packets of any length and content. A packet could thus contain any combination of errors, sources and destinations, and actual data content. Furthermore, network loading could be artificially increased to saturation levels to exercise the node software or develop statistics otherwise unavailable within a specific network situation. Figure 4.62 suggests the types of packets that can be built and the types of tests possible with node simulation. The delay and utilization characteristics of a network under various load conditions is an important consideration in the design and planning stages.

Network Usage Assessments

While an analyzer can chart network performance, it also performs other esoteric duties when used with some imagination. Its functions might include preparing usage statistics for accounting and billing purposes; these statistics could then be used as a basis for charging users and groups for network utilization. Take as an example the information in Figure 4.63. User f in group B is the major user of network resources. Although it might be politically unfeasible to charge group B for his utilization, it might be important to demonstrate demand for network bandwidth resources. It is certainly the case that should the network become saturated and unable to meet the demand for services, network access for user f might need to be changed.

Group C might, for example, be the accounting department that provides the paychecks, payables, and receivables. Network utilization is

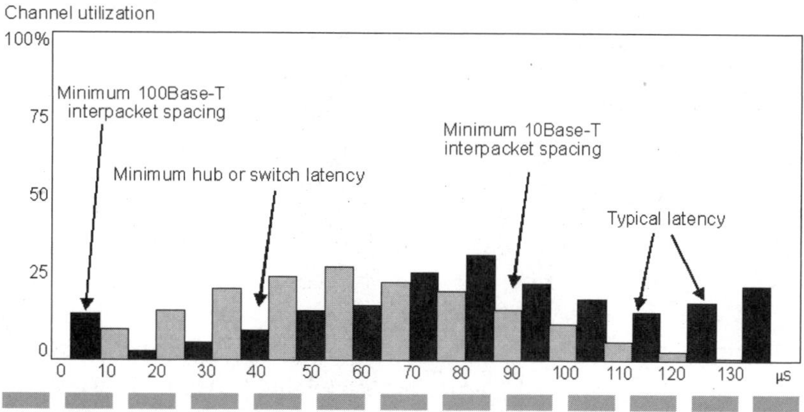

Figure 4.61.
A histogram of network access latency.

Simulation	Anticipated indication
Short packets	Software error handling
Long packets	Software error handling
Long packets	Bridge, switch, router rejection
Corrupted packets	Router rejection
Forced collision packets	Router rejection, NMS alarm
High network loading	Alarms and threshold reports
Misaligned packets	Alarms and high error rates
Source quench messages	Lower traffic levels

Figure 4.62.
Network simulation for testing preconceived theories.

minimal; however, relative value of this work is certainly maximum. This example indicates an implicit resource allocation decision on bandwidth. Although this is a valid use for the analyzer, frankly, in situations where I needed to assign payment for services, I would rely on the NOS accounting functions instead (because it is auditable).

Capture and Store

Another key feature of a network analyzer is the tool's ability to intercept and retain packets. It should be able to capture every packet; although on a busy network, limited available buffer space might prevent any more than seconds or minutes worth of data to be stored. Some analyzers are too slow to capture every packet; they might capture one in five packets.

This is statistically sound if the tool is capable of identifying and tracking each and every packet without storing the packet's contents. Some analyzers are too slow to recognize all packets. This means that packet counts do not accurately represent traffic; this is not sufficient. At a minimum, the statistics must be collected accurately. Node workstations often provide resident software to track this information for that node. Some node workstations provide software for global tracking. If this information is accurate and not just approximate, a protocol analyzer might be unnecessary. The ability to filter selected packets or extract only header information is a powerful tool for statistical compilation, but is usually only available with a protocol analyzer. Also, time stamping of packets is a desirable feature for matching individual packets with their function, as demonstrated in the mail example. It also provides latency and interarrival statistics and a reality check on the statistical accuracy.

Several analyzers store from half a megabyte to four megabytes of packets. They take a photograph of the network activity. On a saturated network, four megabytes might represent only three seconds of activity. At 100-Mbps speeds, protocol analysis requires intelligence to get a few seconds of representative traffic, but mostly cumulative statistics. Captured packets can be stored on disk and analyzed later with word processing tools. For example, four megabytes of network traffic could be stored to disk. A quick view might indicate many mail packets, as in the earlier example. Electronic mail seems to require considerable disk space (confirmed by other tools) and now it seems to require a large portion of the network. How large is a matter for estimate, unless the four megabytes can be analyzed and measured by content. If the file can be edited and parsed, an accurate percentage figure can be compiled. UNIX tools like grep, wc, and awk are especially competent for this analysis; otherwise edits and sorts with line counts are appropriate.

Group	User	Daily average packets	Utilization	Network Utilization	Network Share
A	a	10000	10	0.47	
	b	40000	40	1.88	
	c	50000	50	2.35	
		100000			4.7%
B	d	400000	20	18.94	
	e	700000	35	33.15	
	f	900000	45	42.62	
		2000000			94.7%
C	g	1000	10	0.05	
	h	2000	20	0.09	
	i	7000	70	0.32	
		1000			0.47%

Figure 4.63.
Budgetary allocation by prorated network usage. Packets were captured and collated by source address in order to generate the utilization percentages.

Snooping with the Analyzer

A protocol analyzer—more so than NMS or other software—allows anyone who understands the use of this tool to monitor networks transparently and unobtrusively. By capturing and parsing packets, it is possible to decipher what users are doing on the network, and who is doing things illegally, unethically, or against policy. One use is to see who is using what resources on the network. While this seems Machiavellian, it can prevent unwarranted use of organization resources. It can uncover thefts, breaches in security, or hidden intrusions via unauthorized transceiver taps. Another use of the protocol analyzer, if it has the ability to build specified packets at the bit and byte level, is to falsify packet designs and break whatever security is in place. If this tool can break security for you, this would indicate that others could undermine the security systems as well.

The ability to capture and view packets adds a level of sophistication to network performance tuning. It also promotes spying. Collision counts, packet counts, defective packet counts, and counts by spawning node are information of value and are useful to tuning a network. However, at higher levels of sophistication—when all other problems have been identified and resolved—it becomes important to understand the mix of information actually transmitted. Informed guesses might indicate that a particular set of nodes are used for long-haul uploading and downloading of large files. Other nodes with heavy but sporadic network loads might indicate large electronic mail requirements. By visually inspecting selected packets, it is possible to confirm these hypotheses.

This chapter detailed migration, installation, and testing requirements for copper and fiber Fast Ethernet networks. You learned how to pull wire or fiber, terminate it, test pairs, and deal with the myriad of physical issues that you will necessarily address when retrofitting 10Base-T to 100Base-T4 or 100BaseVG, building a new 100Base-T network in parallel to a preexisting LAN, or reaching other buildings and distant corners with 100Base-FX. In addition, this chapter explained some of the techniques useful for measuring performance of a network experiencing bottlenecks and useful for ensuring that the Fast Ethernet migration achieves the planned goals. The next chapter concludes this book with practical examples of network migrations, including some migration disasters.

CHAPTER 5

Case Examples

This chapter explores some real-world examples of installing Fast Ethernet and migrating from slower local area network protocols, resolving bottlenecked designs, and dealing with the pitfalls of client/server applications. This chapter also includes some of the problems associated with three-tier client/server designs, and how to optimize a network architecture to match the backbone traffic load requirements. This chapter on Fast Ethernet migration strategies combines all the material presented earlier in this book into action plans and actionable migration strategies.

Setting Goals and Priorities

Any network installation, upgrade, or migration should be part of a strategy. You want to set goals and assign priorities to each goal. From there, you plan how to achieve the goals. Sometimes, it is important to define scope, time frame, and scale to your goals before looking at anything else. You may discover that the goals do not mesh with the organization and its requirements, or you may discover that different network migration plans create better results over the long term. Simply to resolve a current problem and not think how it will integrate into future plans is short-sighted; it can also be an unnecessary expense in the greater network plan.

Fast Ethernet is no panacea for all network problems, not even bandwidth bottlenecks. Consider that while bridges and routers are now effective to maintain Ethernet wire speed at 14,400 packets/s, no such devices can interlink Fast Ethernet and maintain the new wire speed of 144,000 packets/s. While Fast Ethernet often does what you expect, you need to be clear about what you expect. Organizational management must also know what to expect. As you will see in some of the cases in this chapter, the Fast Ethernet migration is only part of the bigger picture. For example, the migration may require displacing 200 people for a few weeks while a site is rewired, rebuilt, and refurbished. Fast Ethernet in itself may not address performance bottlenecks, so systems and enterprise communication facilities also require a comparable upgrade. So, if you look at a budget of $400 per node to rewire and migrate 10Base-T to 100Base-T, you may miss another $1500 per user required to enhance the workstations and servers.

Installing or migrating to Fast Ethernet addresses two goals, that of adding more bandwidth, and/or that of decreasing overall network latency. As you saw in chapter 2, migrating to Fast Ethernet may increase network bandwidth at the expense of network latency. You could also migrate to Fast Ethernet and get the new bandwidth in the wrong place. Therefore, you need to create a clear and concise strategy, not just for you, not just for management, but also for the users as well. Their perception of the strategy for the project must mesh with your strategy; one of the most disastrous results of a Fast Ethernet migration is to let users think that 100 Mbps is 10 times faster than 10 Mbps, so that transactions formerly taking 3 to 7 seconds will now occur within 0.3 to 0.7 seconds. It is possible that a network upgrade could yield such stunning improvements, but it really isn't very likely. Although network latency and transmission delays can be a major component of slow processing, it is not the only major component. Communicate the effects, expected results, and the process very clearly.

If you'll recall, in Chapter 2 I calculated the cost for network downtime and also a cost for network slow time. Fast Ethernet migration usually addresses network slow time, not outright network performance failures. If the strategy for a migration is to reduce network crashes, you may want to rethink your goals. Differentiate access from performance requirements and network reliability from application performance requirements. The next example is actually very typical of the types of Fast Ethernet migration requests for proposal that come my way, and demonstrates the strategic misalignment with technical requirements.

Most network hardware, including some very old Ungermann-Bass Net-One (which is about the oldest I have seen running in a mission-critical environment), is reliable even after 10 years. By this I mean it either works or it doesn't. When it doesn't work, you swap out the failed components. The DP manager at an elective-surgery hospital with mostly Net-One equipment approached me to upgrade the network to Fast Ethernet because it was "unreliable" and his programmers and administration people wanted to move to something newer.

The manager was highly experienced with mainframes, not with PCs and networks. He needed first to define goals. First, he wanted to increase the reliability of the network. This perceived problem was solely due to a knowledge deficit, and resulted from some basic administration problems. Fast Ethernet might have solved the initial goal as a byproduct of the upgrade, because Fast Ethernet packet drivers are not available for NetWare 2.02; the migration would have eventually forced some sort of server upgrade in any event, thus addressing the real cause of the network problems. The second goal, that of rationalizing the network infrastructure from a mix of 10Base5 and 10Base2 coaxial cable, 10Base-T, and two different NOSs to support UNIX-based workstations to replace mini-based blood analysis equipment, represented another mismatched strategy.

The minicomputers required 26 hours for certain blood gas analyses and other assays that the UNIX workstations would provide in 20 minutes. The concern was that the results could be delivered without delay to the emergency room. He thought that the current network was inadequate to provide the necessary level of bandwidth because it was localized in the LAN administration area and a few accounting offices.

The manager and I worked out an appropriate strategy to address the goals. The first step to fix the unreliability problems were solved with documentation. The simple solution was merely getting him and his group a photocopy made from out-of-date Net-One manuals so he could get some type of technical support. I also set him up with CompuServe so that he could communicate with other people still running the Ungermann-Bass NOS. I subcontracted to a Novel Platinum Reseller to upgrade from NetWare 2.02 file server to NetWare 3.11; the addition of more server memory solved the "network unreliability" problems, because the servers did not abend so frequently after those changes. I also suggested that he rationalize the network with only NetWare, but he was unwilling to do that immediately.

The second step required determining a strategic plan. Did the hospital want the new blood testing equipment? Could it afford it? In which building would the new lab go? This was not my decision nor the DP administrator's decision, because it depended upon whether the patient area could support a second trauma emergency center and an enhanced on-site laboratory. The issue was more than just that of some blood assaying equipment; other tools also were required at that new site. Issues of whether hospital administration would use E-mail, whether doctors and nurses would write orders on computerized charts, and whether drug orders could be stored electronically also came up. The strategic plan is still under debate by the hospital board of governors; the technical issue of whether the hospital needs Fast Ethernet is irrelevant until some primary business strategies are determined. The strategic plan is also a hostage to budgetary issues that begin with at least $400 per node for the current 30 nodes, and could be as high as $5000 per node for an additional 300 nodes. Strategy ultimately becomes a budgeting issue; this is addressed in the next section.

Think big: The consequences of many decisions may not become apparent for years. New administrators tend to err on the conservative side, finding it difficult to make the leap from their immediate circumstances to the big picture. The game could be lost because of this relative inaction. Often the network administration team will be constrained to repair only those items that show clear failures. While individual problems may be innocuous, a number of marginal problems may combine to produce a global effect. Because the network management role in such a case is a repair, replace, and restore role, forward-thinking action is difficult to achieve. Upper management rarely encourages potentially disruptive risk-taking. Radical changes are frowned on, and network changes are filtered through a conservative screen of consensus. This policy can be the death of the success of a network manager, because growth, technological change, and proliferating problems may undermine that person's authority, perceived competence, and ability to institute large-scale repairs. It can also create a dead-end job. This tendency should be resisted. Not only will conservatism ultimately devalue the network manager and the team, but the organization as an entity may slip behind the competition as new technology supersedes the old.

Conservatism boxes a manager into a suboptimal position and limits the options otherwise available. When decision making is indecisive and risk taking has been avoided as a matter of policy, taking even moderate chances raises eyebrows because others have become accustomed to the norm. Breach of the norm, no matter how risky or risk-averse, will cause objections. Establish your credibility soon. The pattern of risk taking is established early on and success with those risks builds a consistent confidence in the network team. Wishy-washy doesn't work.

Three major factors affect the implementation and usage of a network: reliability, organizational growth, and technological change. These three items determine the level of service achieved and the success of the network administration group. You want to move from survival mode to a strategic role. The solution used by many companies for any network constraint is the addition of more resources. Unfortunately, unlimited DP budgets no longer exist, and network decisions are usually derived from an approach that assumes the necessity of bypassing the stodgy conservatism inherent in legacy host organizations.

On the other hand, networks are incorporating the services formerly provided by telephone systems, and are integrating these services into the production and support activities of the organization. Networks are new resources to most organizations, and despite the many similarities, neither appear as glamorous nor are held in as high esteem as the DP groups were in the 1950s, 1960s, or 1970s. In the 1980s, PCs supplemented overloaded DP departments. Even now, distributed data processing and improved user-layer software have achieved a higher user sophistication, greater user expectations, and less reliance on a DP department. Also, distributed processing power with superservers allows the addition of supposedly less costly computers. They can also be added incrementally rather than as a single, indivisible mainframe. This copes better with trickle growth, and better coordinates needs with resources. Nonetheless, network budgets are modest and more tightly controlled than DP budgets are now. The experience that many organizations have had with uncontrolled DP growth also limits network budgets.

Good network management requires self-imposed financial controls. External controls devalue the esteem of network administration. It is vitally important to justify the expense of every aspect of a network. You cannot add up the numbers and ask management to sign a blank check. You also do not want to go to management and ask for something that is not aligned within organization strategies or requirements. This, of course, speaks to the previously stated need for self-assessment. The intangible benefits of a network do not carry much weight when money can be spent on something else. Check signers want sophisticated cost-benefit analyses. Demonstrate the project payback. Calculate the net present value of the network over time. Show how downsizing DP operations will actually generate money in the present year. Show how the return on the networking project exceeds the organization's cost of capital.

For these reasons, a planning mechanism is necessary to address the existing network and to acquire additional resources to match organizational growth and technological change. Financial planning can be ad hoc in an environment in which there is a cost-plus mentality, or zero-based budgeting. Zero-based is better politically than ad hoc, because every expense has been justified.

Ad hoc planning devalues the decision-making process, because needs are addressed sporadically and without clear controls. It also makes it seem that you are in survival mode. Each newly identified need becomes a battle for budget. Zero-based budgeting schemes are difficult to implement, because they require more management time. All hardware, software, and labor costs are addressed up front, and organizational change becomes a detailed, preplanned financial issue. Growth in head count becomes a known quantity, and technological advancement is no longer an unknown quantity with nebulous ramifications, but rather a clearly stated and clearly controlled financial cost. Figure 5.1 contrasts the advantages of each financial tracking method.

Financial planning encompasses the entire network. Very often resources are allocated to many individuals and groups within the framework of the network. When the network is a "free" resource, like telephone, secretarial services, desk space, or other utilities, the network is devalued. After all, marginal cost is free. When your LAN has remote WAN components, service is not so free. Financial planning provides

Ad hoc	Zero-based
Goal orientation	Nonspecific
Easy	Concise
Fast	Informative
Accurate	Prestigious
Survival mode	Inaccurate (assumptive)
Creates surprises	Time-consuming

Figure 5.1.
Network budgeting choices.

that first step for understanding the resources applied to intercommunication. A budgetary methodology builds that first stage for inventorying network uses. The second step is to gain an understanding of user needs. A network provides services, such as:

- E-mail
- Network fax
- File and print services
- Centralized control
- Centralized management
- Automation
- Remote access to databases, BBSs, and the Internet

Understanding the consumption and depletion of these resources completes the picture of network management. Additional resources certainly will be procured to alleviate network overloads. What types of resources, to which sets of problems these resources will be targeted, and how much will be allocated are, in part, within the purview of the network manager. Strategy ultimately becomes a budgeting issue.

Budgeting

Most technical data communications network books deal with the bits and bytes, not the operational issues. This section covers the unusual topic of budgeting the network installation for a three-floor building, with part of one of those floors sitting below grade in a hillside. The floor plan is the octagonal building illustrated throughout this book. It is a floor plan representative of most commercial office spaces and offices built to suit a corporation. Wiring for the building includes voice and data, and as a network person you should be privy to (and instrumental in) the process of wiring both voice and data. One of the key points of this section is to show relevant and relative project costs, but foremost to pull the cost of a wiring migration project in perspective with other organizational concerns and costs. You should recognize that it is not worthwhile to shave costs on the wiring. Figure 5.2 illustrates the budget for a typical wiring project:

Because the area is 11,056 m² (107,970 ft.²), the communication budget works out to about $102/m² ($10/ft.²). This is consistent with average communication costs at the current time. Because telecommunication equipment typically gets less expensive over time and competitive pressures

Description	Budget	Actual	Changes
IDF equipment	$83,565	$67,065	($13,000)
Voice riser	14,065	10,000	(4,065)
Data riser	6,710	6,710	0
Fiber riser	37,144	37,144	0
Riser tests	2550	2550	0
Basement laterals	37,475	37,475	0
1st floor laterals	51,450	51,450	0
2nd floor laterals	54,600	54,600	0
Panels, plates, jumpers	4,500	4,500	0
Phones (250)	30,000	22,000	(8000)
Servers (15)	186,750	140,000	(46,750)
Workstations (175)	612,500	500,000	(112,000)
NICs (200)	65,000	46,000	(19,000)
Hubs (288 ports)	189,600	150,000	(39,600)
Total	$1,375,909	$1,129,494	($225,350)

Figure 5.2.
Wiring budget for a typical project.

are paring down margins, you should expect to see savings if the budget cycle is far enough advanced over the project deadline. However, the costs for Teflon and other cable are not very competitive, and if anything, you may want to lock in a price ahead of time to prevent material cost overruns.

Some of the savings on servers and workstations also results from vendor price cuts relating to the release of the Pentium as the top-of-the-line desktop and server PC. Some of the savings accrues from making do with 80486 100-MHz servers and a realignment of client processing requirements.

Because the price margin for high-end data communication equipment (such as Fast Ethernet cards and switching hubs) runs around 50 percent, a large order represents both an opportunity for competitive bidding and some tough negotiations with a vendor. If the network design calls for the newest, latest, and greatest, do not expect a discount on equipment.

Because this is a large wiring job, you may be able to get a 30-to-35-percent discount on labor with a smaller local contractor. However, because the labor component of this job runs about $150,000, it may be wiser to go with a national contractor. While they may not offer any financial incentives, you should be able to press for an extended warranty on the cabling infrastructure. Although $37,000 to $50,000 is a lot of money, it usually isn't worth the risk to chase after a local contractor that closes up and fails to complete the job. Even if you sue, what compensation can you expect?

The risk isn't just the cost of the labor and materials (and that cost should be very insignificant if you create a contract with a pay-as-you-go feature), but rather for damages and late fees. Consider the ramifications of a partially completed wiring job with lobes cut too short, installation of subgrade materials, and actual damage to the site. You could be looking at a repeat wiring job. You are risking $150,000 or more to save at most $50,000.

Now, consider the risks when you factor in the building costs (exclusive of land) as well. This structure and its furnishings are likely to cost $7,500,000. The potential savings on a cut-rate wiring job is less than 1 percent of the site costs. Because most building contractors and projects often include a completion bond which costs upwards of 5 percent of the project, saving 1 percent against a 5-percent insurance fee is inane. If you figure that the finance costs for the building run about $12,500 per week, a four-week-long wiring fiasco could erode all possible savings. If you figure that a delay disrupts at least 200 people at a weekly salary in excess of $120,000, you should see that shortcuts on wiring may only come back to haunt you. That the wiring infrastructure is so important is one of the most hidden pitfalls of a migration to Fast Ethernet.

Also consider that the installation process is probably a two-phase process. The risers and lateral wiring are installed after the concrete is poured, but before most of the internal partitions and ceiling are hung. The dress-up with the switch plates and connections usually happens after the walls are painted, while the testing is performed when the building is ready to go live. Typically, problems when wiring the building means that the wires go in after other things are completed; this is more expensive, more damaging, and more disruptive. How much are you willing to risk? You could jeopardize an $8.5-million project on just $200,000 in wiring and $150,000 in labor.

When you start looking at financial budgets of this magnitude, you can understand why many organizations pay an outside consultant $10,000 in advance to design a communication plan and another $20,000 during the course of the project to monitor its progress. So much hinges on getting working phones, working networks, working computer systems, and also a working infrastructure that meets the performance and capacity requirements of the organization.

Performance Analysis

A successful migration to Fast Ethernet means installing a reliable wiring infrastructure, and both chapter 4 and the budgeting section hammered home this point of view. A successful migration to Fast Ethernet also means that the initial strategy was followed. Most migrations are honestly only partially successful because the strategy did not match the technical thinking. At least half the job for a migration is understanding goals and how the technical solutions match the strategy. The goals and analysis section in this chapter addressed that.

A successful migration to Fast Ethernet also means that network bandwidth starvation is appropriately defined and located. The lack of sufficient bandwidth seems to be a critical problem up to about 30 Mbps, then backbone bandwidth becomes a problem or else the base network devices (such as file or database servers) become bottlenecks. It would be nice to grandly state that a duplexed Ethernet or Switch 10 network would be the solution for all these problems, but in reality every network is different. The successful migration requires effective network performance analysis. Refer to *LAN Performance Optimization* (McGraw-Hill

1993), *Computer Performance Optimization* (McGraw-Hill 1994), or *Enterprise Network Performance Optimization* (McGraw-Hill 1995) for all the details on the tools and techniques of network and system performance analysis.

The rest of this chapter is about how to pinpoint bandwidth starvation and how to address it. The next figure (Figure 5.3) is a network flow diagram that shows current loads and then forecasts potential loading on the network. It is not to be confused with a physical diagram or a logical diagram.

The process of gathering information on current sustainable traffic loads shows that this client/server network is running at capacity with around 9.2 Mbps. The current configuration is deferring an aggregate peak load that could reach 33 Mbps, but would more likely be sustained at 16 Mbps if the Ethernet channel could somehow support that. If you aggregate the highest possible network traffic with this type of configuration, the backbone could see peak traffic loads at 186 Mbps. More likely,

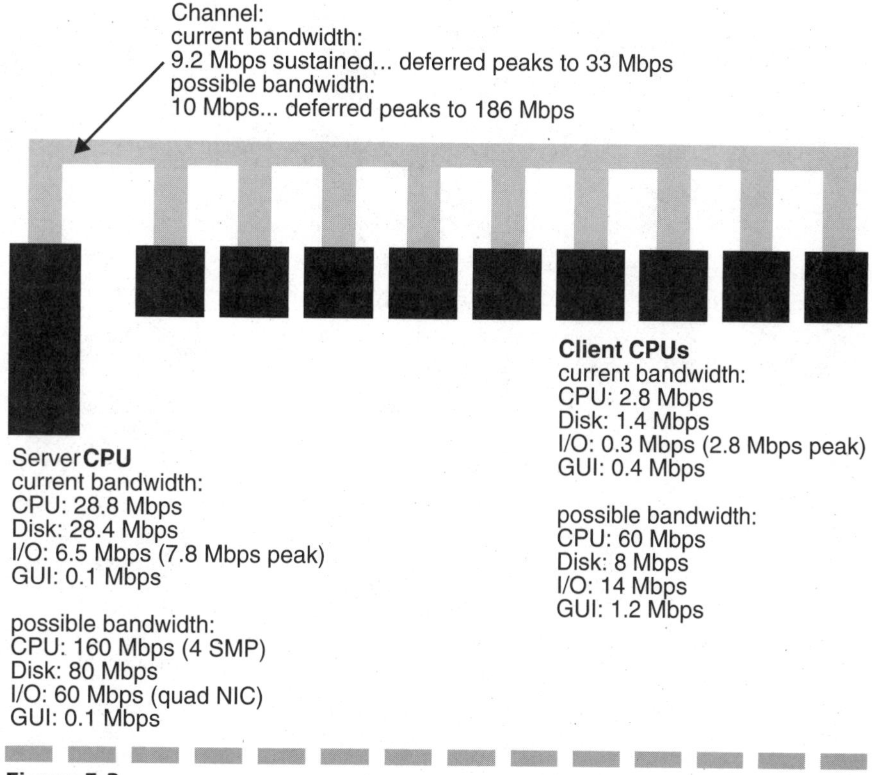

Figure 5.3.
Network traffic flow diagram.

the hardware could sustain a traffic level at around 90 Mbps. The question is, can the user applications sustain that level, or will it more likely fall somewhere between 16 to 33 Mbps with an enhanced network channel? The answer is found by extrapolating user traffic for the client/server application for one user to all nine users. If you have a larger network, just scale up the loads by the number of users that you have.

Although network traffic is inherently exponential and any problems on the network channel create exponential delays, the user work is mostly arithmetic. If each user transaction represents 1 MB, nine comparable transactions will generate 9 MB. If each transaction can occur in a second, there is nothing to guarantee that nine such transactions can occur in the same second. So, if the network traffic for one user is 0.3 Mbps, the network traffic for nine of the same users will be more than 3 Mbps, perhaps as high as 33 Mbps (as the prior diagram suggests). User loads scale, but network traffic loads do not. That is the difficulty with most performance analysis performed to justify Fast Ethernet migration. You have to factor in the protocol effects, timing issues, and queueing delays.

Migrating the Server(s)

The server in the prior diagram created 20 times the network traffic of each user device. Asymmetric loading is very typical in networks supporting a file or database server, or a network specifically supporting a client/server application. All user nodes can usually coexist on a network channel, even if its capacity is only 10 Mbps, but the traffic from the server causes most of the bottleneck. If you can microsegment the network so that server talks to different segments, you can often resolve many bandwidth blues. In fact, at least six different vendors create quad-port Ethernet cards so that a server can talk to clients on four different segments. This often addresses client/server bandwidth problems without any new wiring or any need for Fast Ethernet.

However, when multiple servers exist on the same backbone and talk to a mix of clients as workgroups evolve, the problem is more complex. If you split the available servers so that each is on its own 100-Mbps link to that backplane, it is unlikely the aggregate can saturate the Fast Ethernet hub. However, if the traffic levels are like those profiled in the prior diagram, you still have the same loads going out on the other links, as Figure 5.4 shows, and the problem remains unsolved.

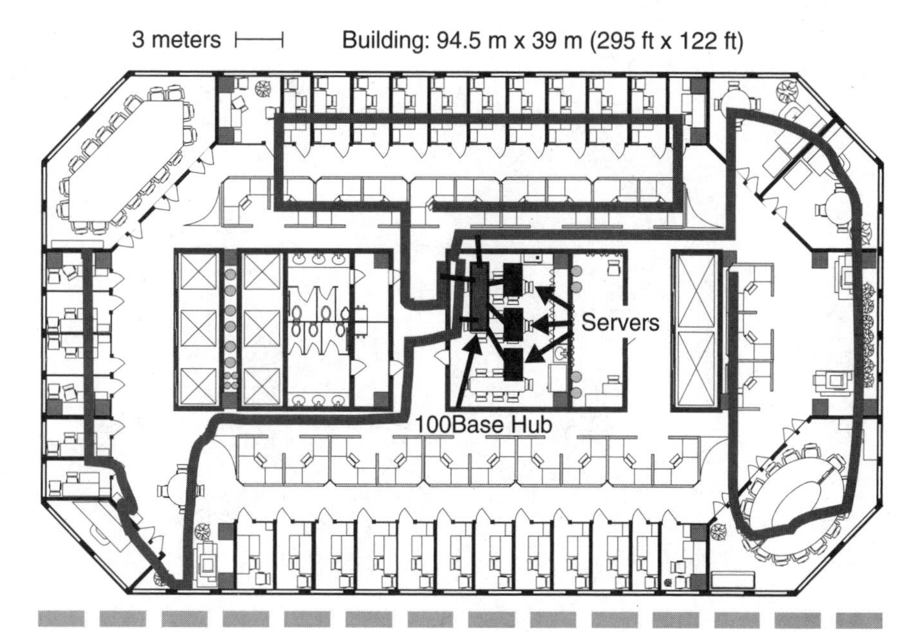

Figure 5.4.
Adding a high-speed backplane for servers with Fast Ethernet gives the servers somewhere to transmit, but the client/server bottleneck still exists on the network, and the same bottleneck persists on the network.

If you could replace the Ethernet backbone with Fast Ethernet, you could address the server load limitations. That means rewiring the network with twisted-pair; for 10Base-T in fact, with potential support for 100Base-T. The aggregate loads are such that you want the servers on a Fast Ethernet hub, but the loads to each client are low enough for 10Base-T to suffice. The wiring design shows foresight if you run at least four pairs of Category 5 to each desk. The mixed 10/100Base hub provides primitive traffic management sufficient to resolve the bandwidth limitations, as shown in Figure 5.5.

The key to this architecture is that you can microsegment the 10Base-T in different layouts to keep the traffic at acceptable levels on each segment. However, 10Base-T still echoes the Ethernet bus layout so that you could have bandwidth bottlenecks. Clearly, not all client/server bottlenecks are addressed by a mixed 10-/100-Mbps wiring scheme. It is possible that the hub cannot support the bandwidth of the outgoing server traffic. It is also possible that the new segments will not support the traffic loads. The solution is to increase the microsegmentation by replacing the hub with a switch. It might not be necessary, and it is

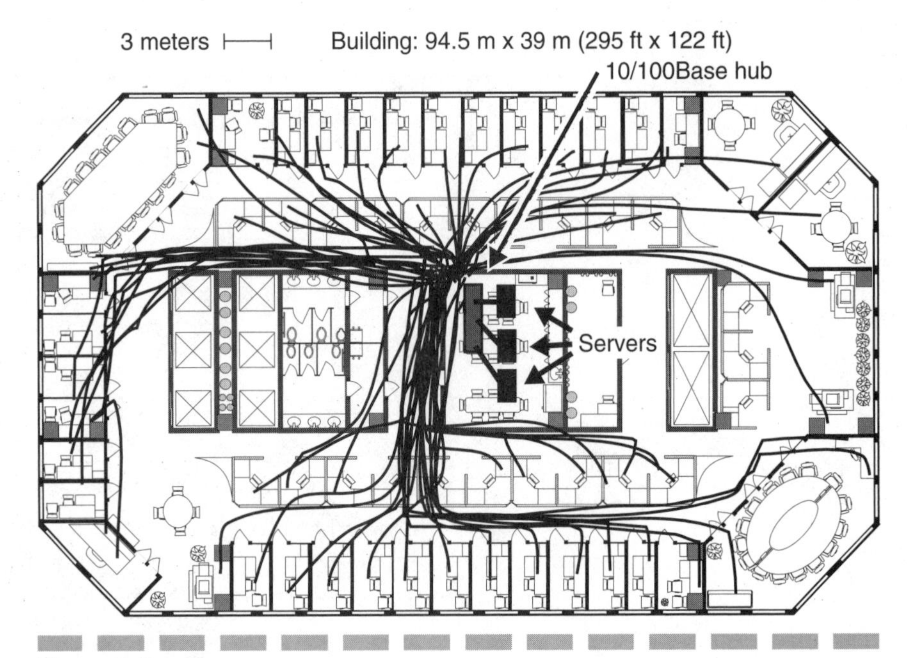

Figure 5.5.
The client/server bandwidth solution begins with at least 10Base-T connectivity, 100Base-T server access, and backbone traffic management.

usually unlikely, that every client node would need its own switched segment. Usually, each switched segment can support from 10 to 20 client nodes. Each server would attach to a switched 100-Mbps port. Ultimately, this configuration provides three additional levels for bandwidth growth.

At the first level, you could migrate individual client nodes to a fully-switched 10Base architecture. This way, each client gets 10 Mbps of bandwidth. The wiring and the current NIC for each client is sufficient; you only invest in more switched 10-Mbps ports. Because most client/server applications are unlikely to create 10-Mbps streams from a server to each node (and this would overload the server, even a UNIX superserver) anyway, this migration is a good bet.

Bandwidth-intensive database, CAD, GIS, graphics, or imaging operations conceivably can outstrip 10-Mbps bandwidth in peaks. A typical four-color image file for a magazine is from 50 to 100 MB in length. Because the image is shipped from the file server for color correction, manipulation, layering, and other operations usually in its entirety, such bandwidth requirements are really infrequent; the image is fetched once and stored once for each production cycle. To address peak loads consistent with this type of processing, the next level is to replace the 10Base-T node links with 100Base-T links to every client node. Realistically, you will also need to address the file server, multi-tier, or replicated database structures so that they can sustain the file activity presumed to travel over the network. A NetWare file server is not the answer. A host mainframe is not the answer. A multiprocessing server with a multiple buses might begin to address the file traffic. Because the bandwidth requirements are sporadic, a shared-media architecture is sufficient to start. Microsegment the clients with some intelligence so that you do not recreate the initial client/server bottleneck. This upgrade requires replacement of client NICs and hubs supporting 100Base-T. The upgrade can be partial or complete, as organizational finances and performance requirements dictate.

The third level requires the exchange of 100Base-T hub for a 100Base-T switched hub. Migrate the overloads to 100-Mbps switched ports just as the servers were initially moved to switched 100-Mbps ports. The upgrade requires replacing some 100Base-T ports with switched ports. As before, the upgrade can be partial or complete, as organizational finances and performance requirements dictate. Again, few activities demand such bandwidths, and few devices can sustain the I/O loads across file systems and system buses. At this level of switched 100-Mbps service for both client and server nodes, you are likely to be building a network with sufficient capacity to meet your needs for the next five years.

Integrating Mixed Speeds

Although the last section explained several effective and partial Ethernet migration scenarios, there is one more scenario that addresses a common bandwidth requirement. Software development groups or imaging groups may need more bandwidth than most other groups of users. While the following figure shows this group concentrated within a small area, premise wiring usually handles any distribution of high-bandwidth users (See Figure 5.6).

The primary benefit of this layout is that Fast Ethernet could represent a quick, temporary, or trial installation. You could run the sixteen lines along the ceiling or floor and pull a single cable through the ceiling to the current network computer room. One Fast Ethernet hub (it is shown as 100Base-T, but it could be 100Base-T4, 100BaseVG, or a switched solution), serves the new network. A second hub, although it

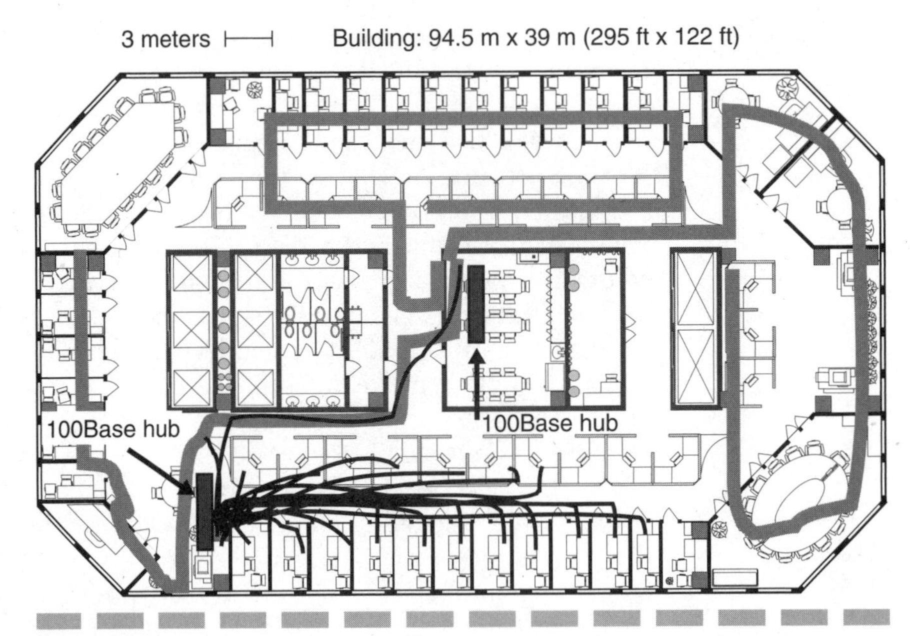

Figure 5.6.
Partial migration to Fast Ethernet for high-bandwidth users.

could be a media bridge, provides a connection to all the servers and all the standard coaxial Ethernet segments. You could install only a single hub at the site of the second hub, but that would probably require a permanent commitment to rewire for Category 5. The cafeteria may not be the place you want to install punchdown blocks, patch panels, and the Fast Ethernet equipment. This issue is addressed later in the wiring closet organization section.

Token-Ring and VG-AnyLAN

Hewlett-Packard, ODS, and Madge have finally released VG-AnyLAN support for Token-Ring. This means that 4-Mbps or 16-Mbps Token-Rings with bottlenecks on the channel can be migrated to Fast Ethernet. Bandwidth utilization is not the measure to use for determining a bottleneck. Instead, use token rotation time (TRT) as the measure. Anything above 65 ms represents a good candidate for migration, because you are going to trade that latency for 20 to 40 ms delay at the hub. Figure 5.7 shows the ring configuration for Token-Ring.

However, Token-Ring is actually wired as a physical star. The standard shielded cable for IBM Token-Ring is IBM Type 1 or IBM Type 2, was installed with lobe lengths less than 250 m (and more likely less than 200 m to meet ring-wrap requirements), and can pass a Category 3 scanner test. In all likelihood, this cable will also pass a Category 5 test and can support bandwidths to at least 155 MHz. The actual physical wiring layout is illustrated in Figure 5.8.

That looks so convenient for migration to 100BaseVG-AnyLAN, except for one serious problem. If you recall from chapter 4, Token-Ring only requires two pairs (pairs 1 and 2), much like 10Base-T, and the IBM cable has only four wires. The wire is not really paired, but it is twisted and it also is shielded with a metal foil or screen. Unless someone was very insightful and installed four or more pairs of UTP, you will likely have to recable. If multiple data and phone ports were installed to each office switch plate, there may be spares that test out to Category 3. Figure 5.9 shows that if four pairs are available, you can migrate from Token-Ring merely by replacing the unpowered MAUs or the powered hubs with a

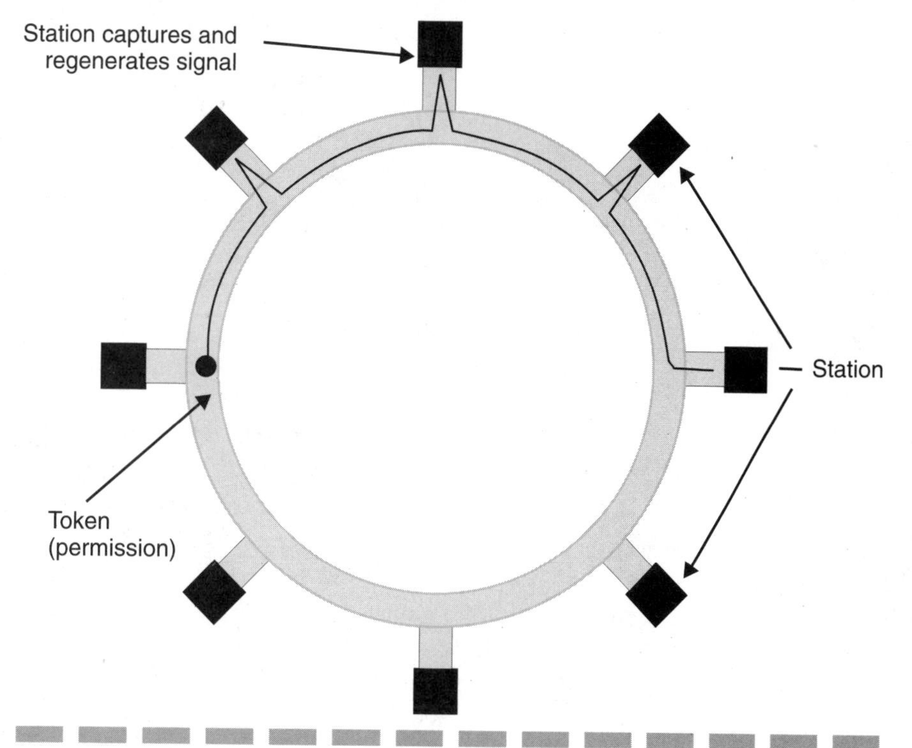

Figure 5.7.
Token-Ring logical layout.

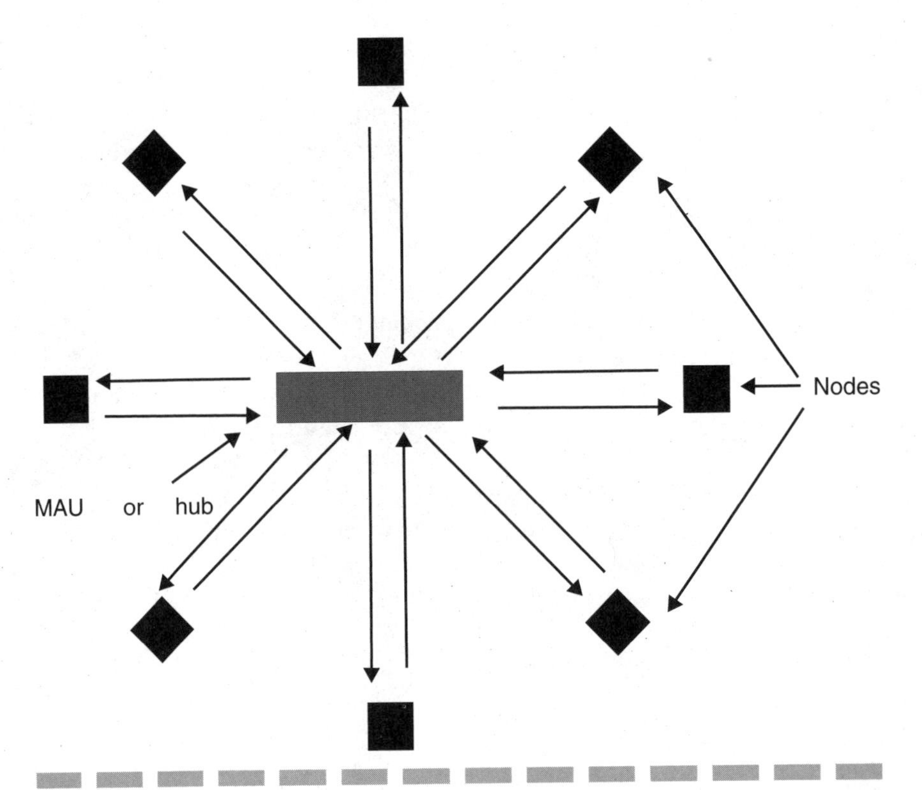

Figure 5.8.
Token-Ring physical wiring layout.

100BaseVG hub. You will also have to replace Token-Ring network adapters with 100BaseVG adapters.

When four pairs are not available and there is no reason to retain the Token-Ring protocol (or mix Ethernet and Token-Ring), consider migrating completely to Fast Ethernet. The Token-Ring wiring will support 100Base-T. You will need to change the hubs, the NICs, and the software driver support from Token-Ring to Ethernet. In fact, because some 50 percent of all Token-Ring networks run Novel NetWare, the switchover should be as simple as rebinding the IPX/SPX client kernel for 1500-byte packet support rather than the 4032-byte packets supported with Token-Ring. It really is not that difficult. In addition, because patch panels and switch plates probably had RJ-45 or genderless connectors (which are not directly compatible with the Fast Ethernet connectorization), the easiest solution is to build custom jumpers to match up the pinout. Refer to chapter 4 for the USOC and EIA/TIA two-pair wiring pinouts.

Rewiring the Floor

If none of the existing wiring systems provide a simple migration path to Fast Ethernet, you will have to rewire. Perhaps the most important decision you will make is where to create the new wiring closet. Placement raises issues of maintenance, proximity to phone PBXs or external WAN lines, and security. Also, this migration usually brings in equipment that did not exist in any other form before. The cafeteria is not the place for wiring systems, hubs, and servers even if it is centralized on the floor. Although it is possible to move the cafeteria or partition a new room within the cafeteria to make space for a server and wiring closet, the next figure shows three possible places for a professional wiring closet. Figure 5.10 illustrates these positions and the extreme wire runs.

The cubicles on the far left actually stretch the Category 3 and Category 5 basic link to the maximum 90 m. You might get away with this position, but there is only a limited margin for error. The equipment room next to the cafeteria is the optimum position, because there are

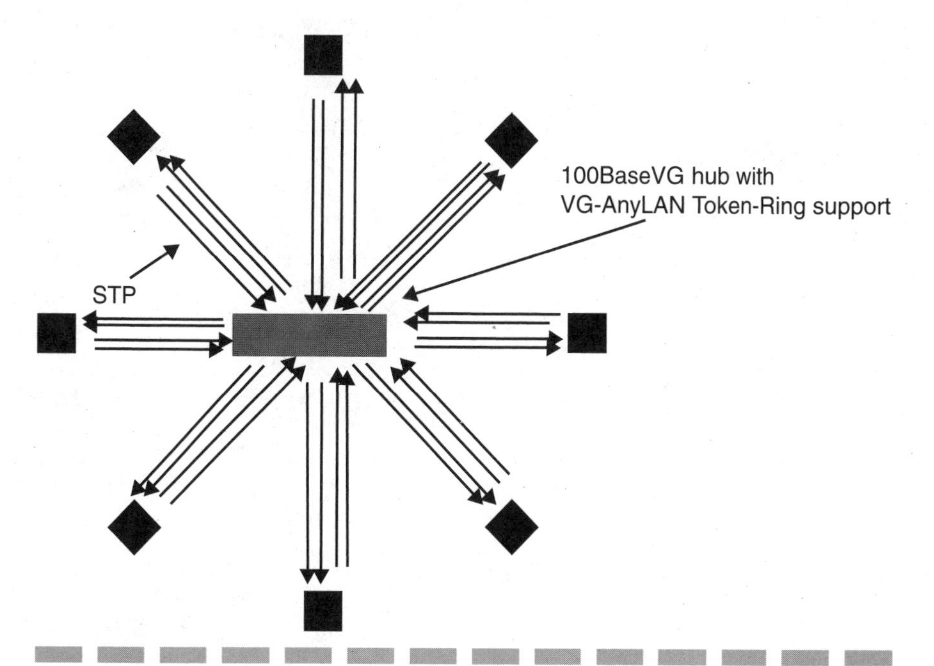

Figure 5.9.
With four pairs of Category 3 wire available, migration to 100BaseVG is as simple as replacing node NICs and the hub.

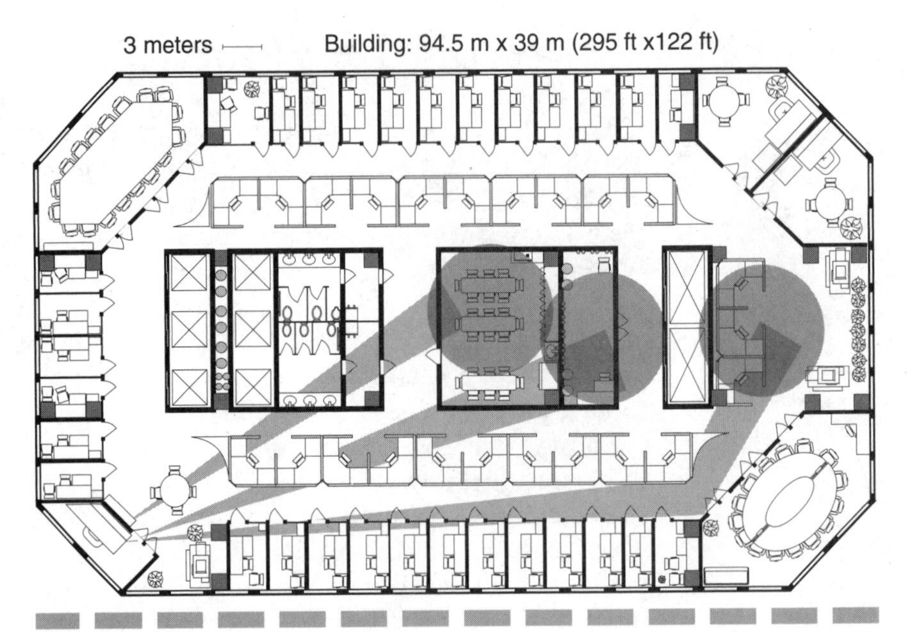

Figure 5.10.
Possible positions for a Category 5 wiring closet and the extreme wire runs.

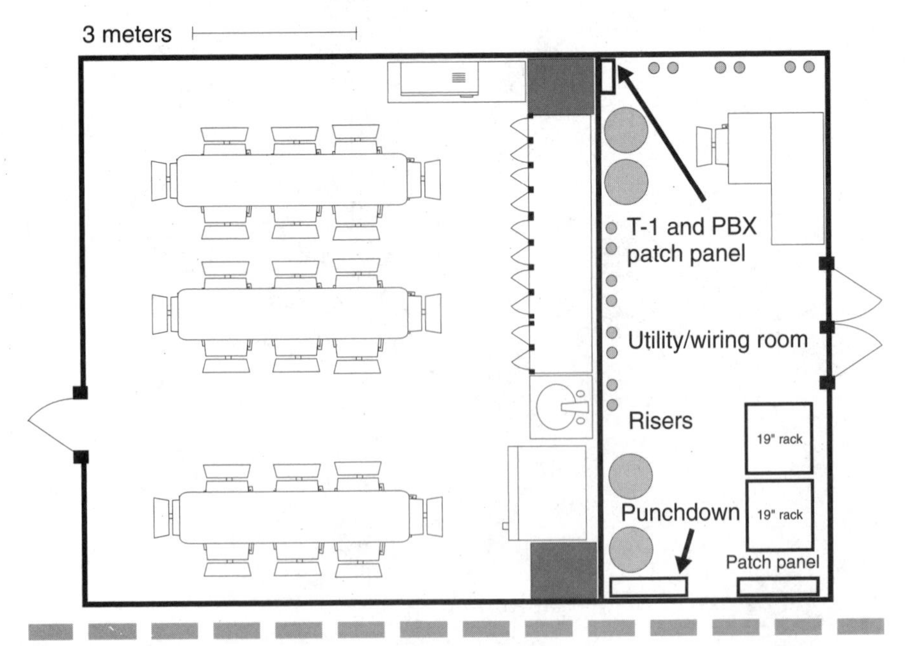

Figure 5.11.
Wiring the equipment room.

conduits in place for risers, there is spare and built-in HVAC capacity for equipment, and the phones and the WAN connections are already there. If you rearrange the furniture and get the desk out of the corner, there is room for several panels and industry-standard 19" racks. Most enterprise hubs and many smaller network devices bolt into these racks. They are useful because they roll (install casters), they provide space for wiring, and wire combs can be attached to organize all the wiring that ends on those racks. These racks cost about $400 to $1000 new depending on security, power, and other options, but are available for $50 to $100 used because they were the standard mounting for obsolete reel backup tapes, big disks, and minicomputers. Figure 5.11 shows how this equipment room might be used.

Between all the risers and vertical pipes for water and sewer, there is plenty of room for servers, printers, and racks for backup media. Because there are often open ceilings (in other words, no false hanging ceiling), access for installing lateral wiring is simplified. The next figure, Figure 5.12, shows the basic paths for most of the new wiring. Typically, for a layout like this the installers will pull a large breakout cable with 100 pairs along the corridors and then drop four pairs or eight pairs in each office or cubicle. Alternately, the installers will pull 25 sets of four-,

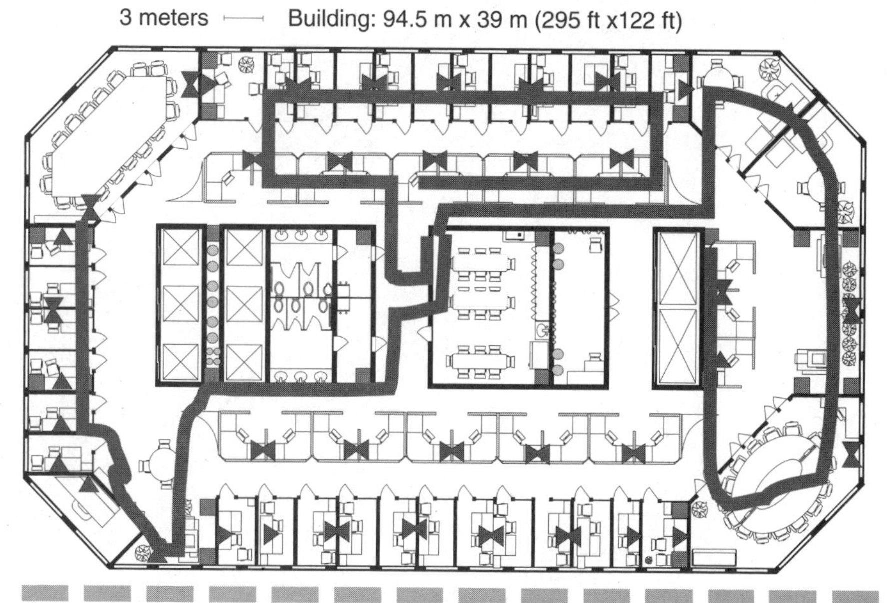

Figure 5.12.
Cable runs from the wiring closet.

eight-, or twelve-pair cables along the corridors at one time and then pull each individual cable to an office or cubicle. A wire caddie holds that many rolls of cable and lets it unroll as it get pulled through the ceiling or conduits. The advantage is obvious both in terms of labor spent pulling cable initially, but also in terms of securing the bunches of cable along that route.

If the transition is from 10Base5 to 10/100Base-T, creating a new wiring closet and equipment room also makes a lot of sense. The transition is probably performed so that the existing network remains fully functional until the new network is complete. Cable pulling takes at least a few days and few businesses tolerate that downtime. I do not like cable installers working near my servers. It doesn't take much to trip over any one of the many wires lying around the floor, dislodging a software security dongle or knocking some piece of equipment over. That brings the network down quickly and adds to user frustration, expectations, and irritability. Drag a couple of unsecured cables from the old site to the new site for interconnections and testing, as shown in Figure 5.13.

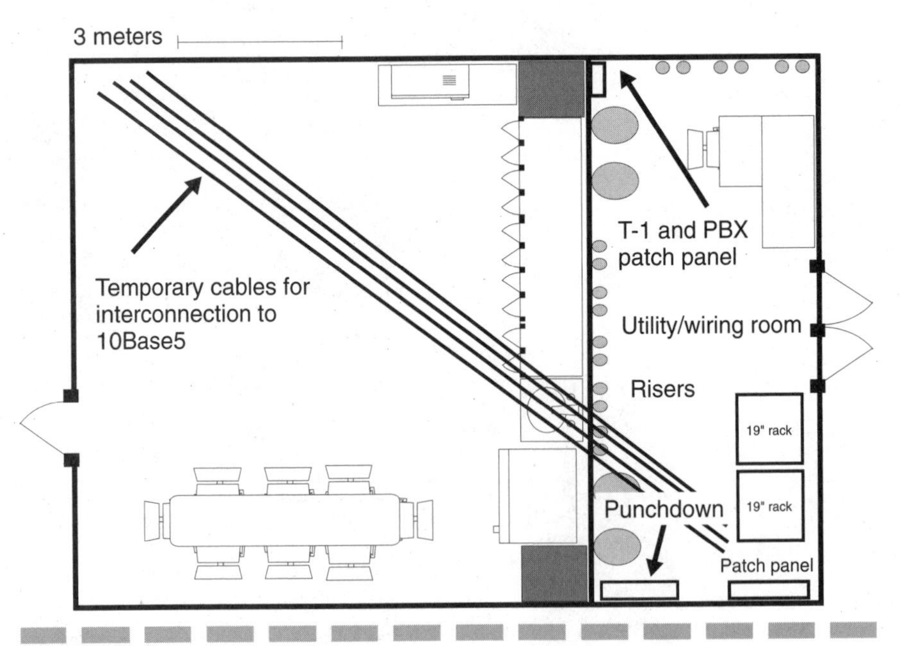

Figure 5.13.
Jumper the new network with the old on a temporary basis.

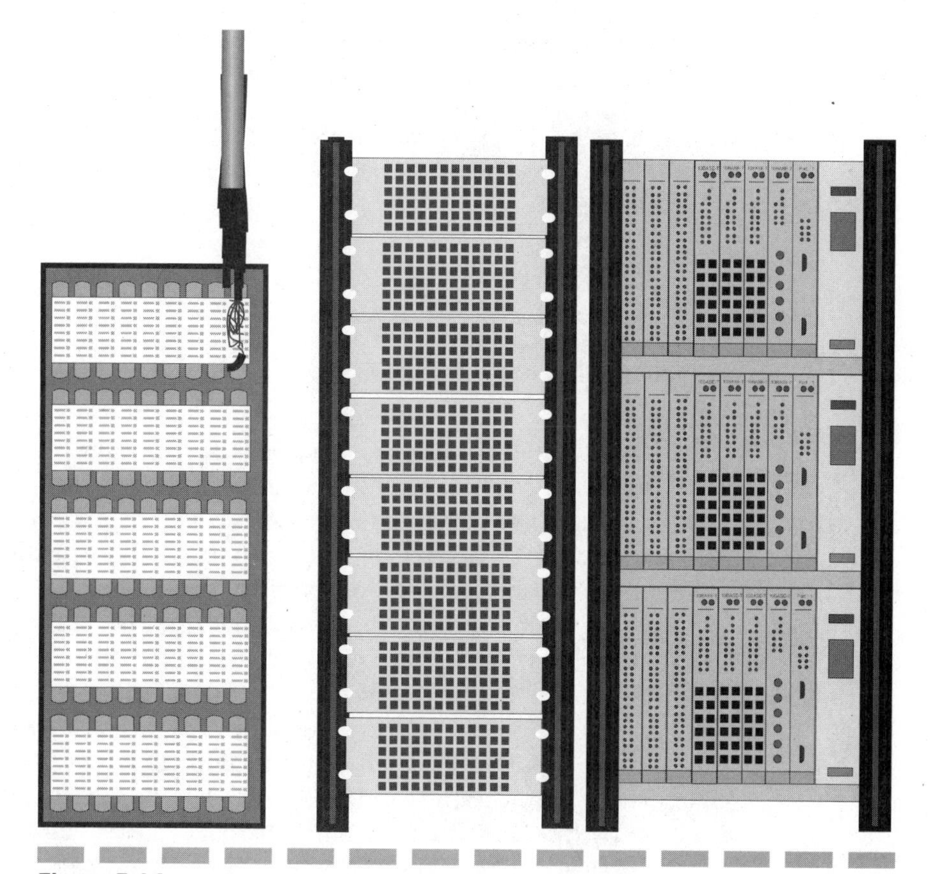

Figure 5.14.
The organization of an orderly wiring closet improves the efficiency of wiring changes while also simplifying any critical wiring repairs.

Wiring Closet Organization

The benefits and rational for premise or structured wiring is best shown in Figure 5.14.

As you can see, the lateral and vertical wiring comes in to a punch-down block (Type 66 or 110) on the left. Typically, these blocks are attached directly to the wall or screwed into a raised wooden panel. From there a hydra or individual patch cords connect into the patch panel in the middle rack. It is possible for these hydras or inbound patch cords to connect into the back of the patch panels. In fact, this improves

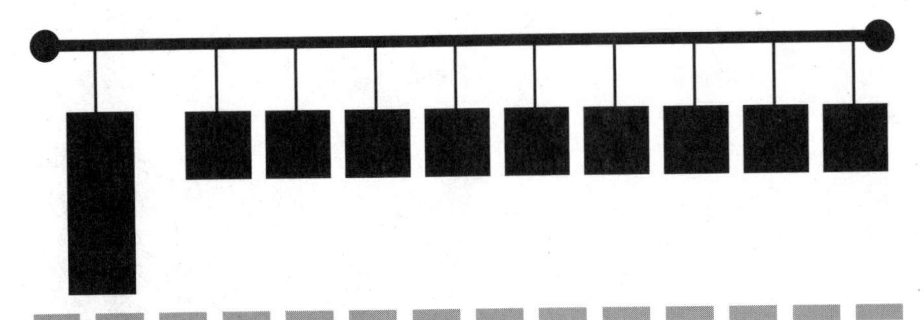

Figure 5.15.
A typical bandwidth limitation in a LAN.

the overall order of the wiring system. Individual jumpers are connected from the patch panel into the network wiring hubs. If (or rather when) a line fails, you can swap jumpers, locations, or ports without the need for any tools to isolate the failed component. You can tell whether the port, the patch, or the lobe to an office or cubicle failed.

If you notice, both the patch panels and network hubs are installed in standard 19" racks. In actual use, the racks will be nearly hidden by the layers of wire that connect into them. Once the racks are fully wired, you really cannot move them; jumpers are limited to about 3 m. However, you can pull racks out and around enough to squeeze in behind them to check connections, fuses, or power supplies. You can also pull out an entire panel or chassis that may have failed without having to restack anything above or below that unit. It is not really easy moving a unit when several hundred jumpers each connect into chassis above and below, but it is less disruptive than trying to adjust a problem with stable hubs.

Resolving the Bandwidth Woes

The most typical situation appropriate for migration to Fast Ethernet is represented by the logical diagram in Figure 5.15. This LAN, although the network could be a microsegmented LAN interconnected by bridges and routers, usually has some number of central workgroup servers and clients running standard office applications. Corporate offices run those applications along with connectivity to a host-based accounting system. Managers who are aware of the strategic value of information often create specialized computer telephony applications and integrate them into the workflow. By the way, this is one reason to support collocation of telecommunication and data communication wiring. Legal offices often run Word for Windows in conjunction with a document tracking and management system such as PC-DOCs. You might note that this logical network diagram in Figure 5.15 corresponds to the traffic flow diagram in Figure 5.3.

The primary solution to a backbone bandwidth bottleneck is to increase the capacity of the backbone. Although the logical structure for any Ethernet is fundamentally a bus, you can increase the bandwidth by collapsing the bus into a hub, switch, enterprise chassis. Figure 5.16 shows the ineffectual results of migrating from 10Base5 to 10Base-T without any other changes.

The problem is that the connection primarily to the server is still limited to 10 Mbps. Although the backbone could support from 400 Mbps to 10 Gbps depending on the selection of the hub or chassis, no fundamental improvements have been made to widen the communication path to the server. When the migration provides increased bandwidth between server, client nodes, and the collapsed backbone, then you are effectively providing increased network bandwidth and addressing the network bottleneck. If the hub in Figure 5.16 provides a partial or complete switched 10-Mbps connection to server and client nodes or a shared 100 Mbps environment, you are addressing the bandwidth bottlenecks of the original network. In fact, duplexed support, particularly for the server, will resolve the immediate bandwidth bottleneck, as shown by Figure 5.17. This is useful for an immediate and simple migration to address the client/server asymmetric overloading.

The next illustration mixes the logical and physical diagrams. In this case, the generic premise wiring system creates opportunities for various

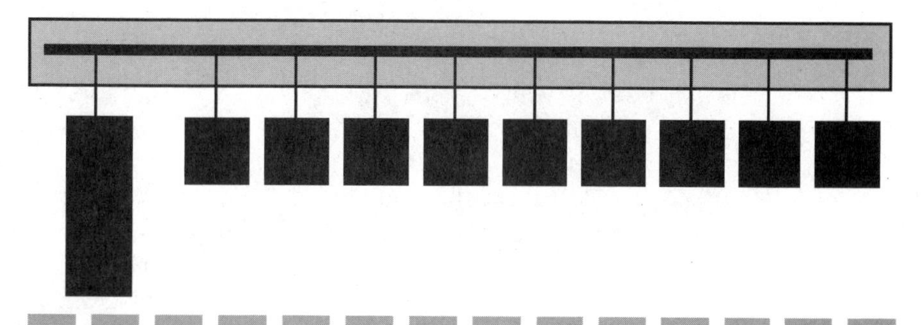

Figure 5.16.
Migration from a bus-based LAN to UTP and a backbone provides only a physical wiring change but no bandwidth enhancement.

Figure 5.17.
A duplexed connection, mostly to the server, will resolve file server I/O limitations.

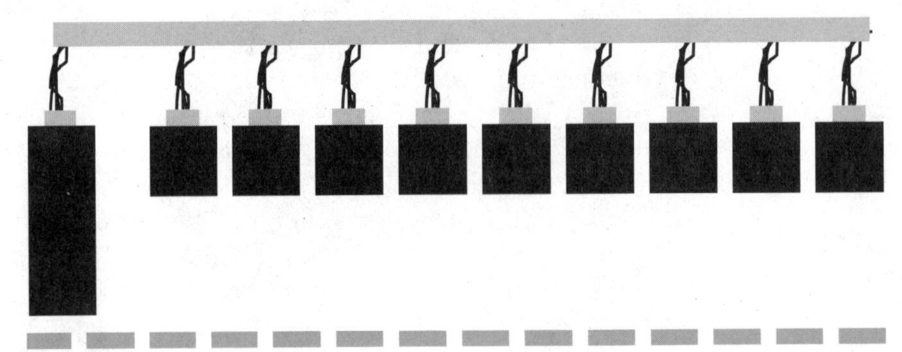

Figure 5.18.
The logical layout specifically shown with a twisted-pair infrastructure.

configurations. Figure 5.18 shows the network of 5.15 through 5.17 rewired with at least four pairs of Category 5 cable.

This infrastructure is important because we can really begin to address incremental network redesigns. Typically, migration to Fast Ethernet is not a total conversion. First, it is not always clear where the bottlenecks are (even if you have statistics gathered from protocol analysis matched with performance information from a code profiler) and second, finances usually drive an incremental migration. However, for an incremental migration to be possible, you need the premise wiring structure. The prior figure could indicate a 10Base-T, a 100Base-T, 100BaseVG, a switched 10, a duplexed 10, a switched 100, or switched and duplexed 100 configuration. There is nothing to define or limit that diagram to any one configuration. That network could even be a Token-Ring or an ATM network.

The value of this diagramming becomes apparent when the network is not just one server and nine client nodes. The network could be extended to 18 clients, or each client node on the logical diagram could represent 10, 20, or 100 nodes. In fact, the logical diagram is the perfect match for extensible and flexible premise wiring systems, as shown by Figure 5.19.

This case shows how important it is to think in terms of logical diagrams apart from physical ones. Solutions to bandwidth woes are usually not complete with the first iteration unless the network designer has very good performance analysis skills (so that current, deferred, and forecast workloads and characteristics of those workloads are accurately defined from the beginning). More often, bandwidth enhancement is a progressive event of enhancements. For example, consider the enterprise solution in Figure 5.20; it provides switched access for all nodes.

You should be aware, however, that Figure 5.20 could be representative only of the client/server architecture, but not of the quantity of equipment. Sometimes the details obscure the important points. If in fact the environment consists of 50 servers and 100 workgroups, it is impossible to link every server into every network because of the number of connections needed. Furthermore, that structure will not sustain the traffic with any current backbone, physical or collapsed. Instead, you will need to create some layers interconnected by bridges, routers, or switches, as shown in Figure 5.21.

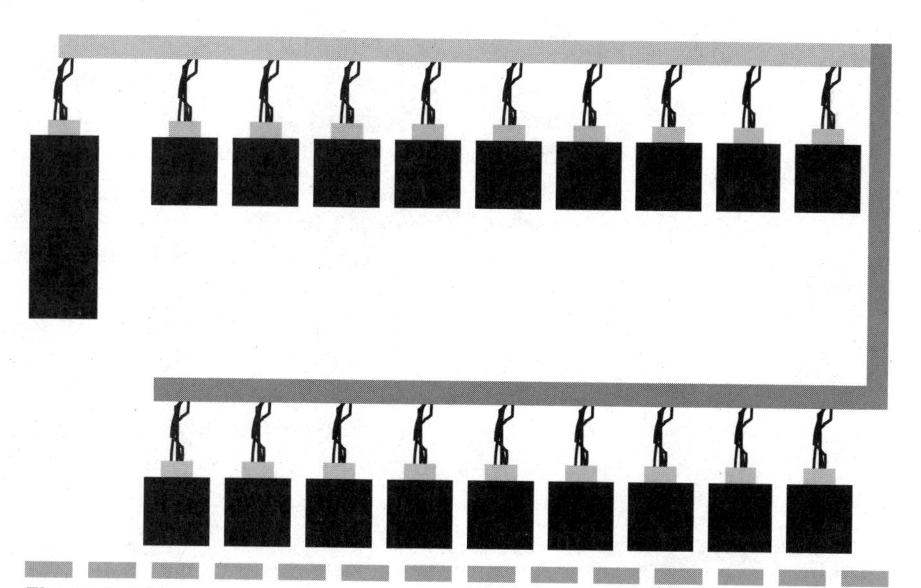

Figure 5.19.
Logical diagrams are extensible, just as with premise wiring systems.

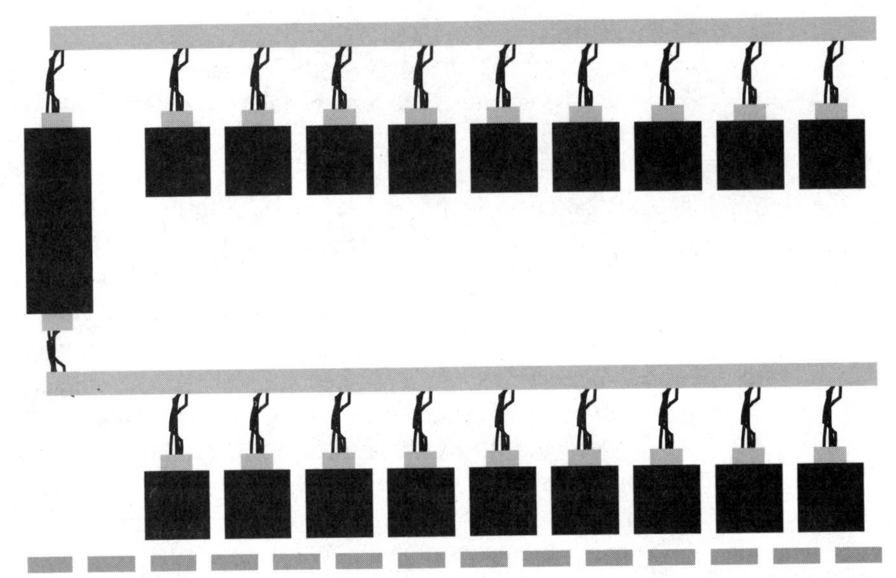

Figure 5.20.
Workgroup hubs connected to an enterprise solution.

Although Fast Ethernet provides ten times the bandwidth of Ethernet and eight times that of Token-Ring, 100 Mbps is still a fundamental limitation for links from workgroups to a backbone. Microsegmentation, which was the first technique to relieve bandwidth bottlenecks, is still very much an important tool as LANs become networks of LANs, the traffic between LANs increases, and facilities are centralized. Although switched 100-Mbps and switched duplexed 100-Mbps connections provide quite of bit of bandwidth headroom for the time being for most sites, there will be a need for Ultra Fast Ethernet with its 1 Gbps to link workgroups to the enterprise hub. Until then, microsegmentation provides a tradeoff in terms of bandwidth for latency. As the prior illustration shows, you will need to keep the number of levels in the cascade to a minimum, because even with switching times approaching 10 to 20 ms with some of the new equipment, these latencies still accumulate quickly to create significant delays with client/server, MS Windows OLE automation, and replicated or remote functions. Vendors are increasing the performance of enterprise hubs to support two-layer architectures as diagrammed in Figure 5.21, so that 10-Gbps and switched 100-Mbps service to workgroups or hubs on each floor of building can sustain the bandwidth required for local applications, the traffic between servers and groups, and the traffic among groups themselves.

The final example for this book shows the two-phased step an international organization made to empower workers with information over six years. Clearly, Fast Ethernet migrations do not often happen in an extraordinary night or weekend of frenetic activity. Rather, they are journeys that parallel the growth in technical sophistication and the integration of workflows into network activities. This Fortune 1000 organization in the transportation business has about 25,000 employees worldwide, with approximately 5000 people sited at this corporate headquarters. Although information had been a mainstay for the rapid growth and strategic expansion of this corporation, a subtle shift was occurring whereby employees were realizing that they were selling information in a timely manner and that the actual product had become incidental. The differentiation between this company and its competitors was in the supplemental value of the information. As such, the information group decided to extend the network infrastructure from serving occasional users to complete connectivity. Workgroup networks had already been consolidated to a central network server facility on the second floor. Even though the building contained a large atrium, 10Base5 was sufficient to reach all existing networked workgroups.

However, as user count increased, server count increased, and the need for fanout boxes to support the increased user density increased.

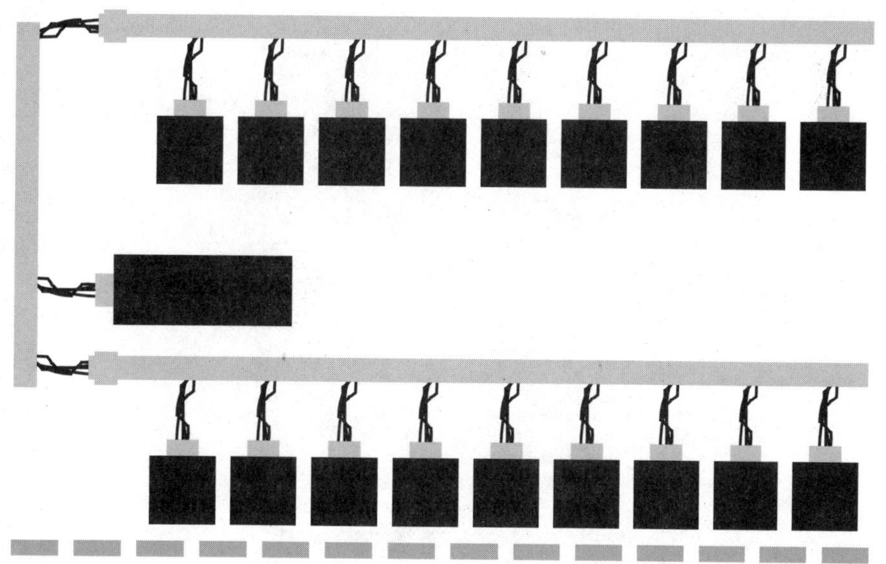

Figure 5.21.
A dual-attached server connected to workgroup hubs.

In the late 1980s, 10Base-T hubs were installed in place of fanout units, but these were still connected to 10Base5 backbones. As traffic loads increased, new backbones were added, and more servers were added. The fundamental structure for each wing (and each floor contained about 16 such wings) is shown in Figure 5.22.

As is typical with networked LAN growth, reliability suffered and user growth overloaded the backbones. Prior attempts to rationalize the many networks were poorly funded, and were stopgap solutions at best. The company sought a unified solution. A massive restructuring with layoffs provided the first freedom to rebuild the network. Several wings were totally depopulated in the reorganization. This provided the working space to strip out the old phone system and its wiring, refurbish the offices and cubicles, and rewire the wings for advanced phone systems and data communications. The technology available at the time was 10Base-T and serial lines to AS/400s. Central wiring closets were created to support this wiring migration (rather than just the central facility on the second floor), and a complete rewiring was completed for about 10 wings. No wiring standards were used, although all pairs were qualified for support of 10Base-T. The wire was not qualified as Category 3,

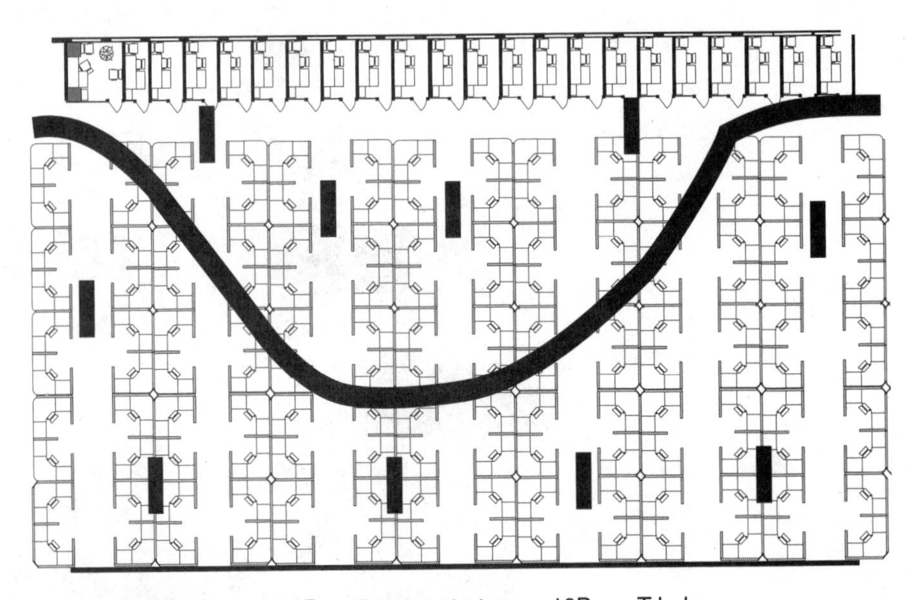

Figure 5.22.
The initial 10Base5 layout for each wing. 10Base5 concentrators or 10Base-T hubs were wired into the ceiling to support high user densities.

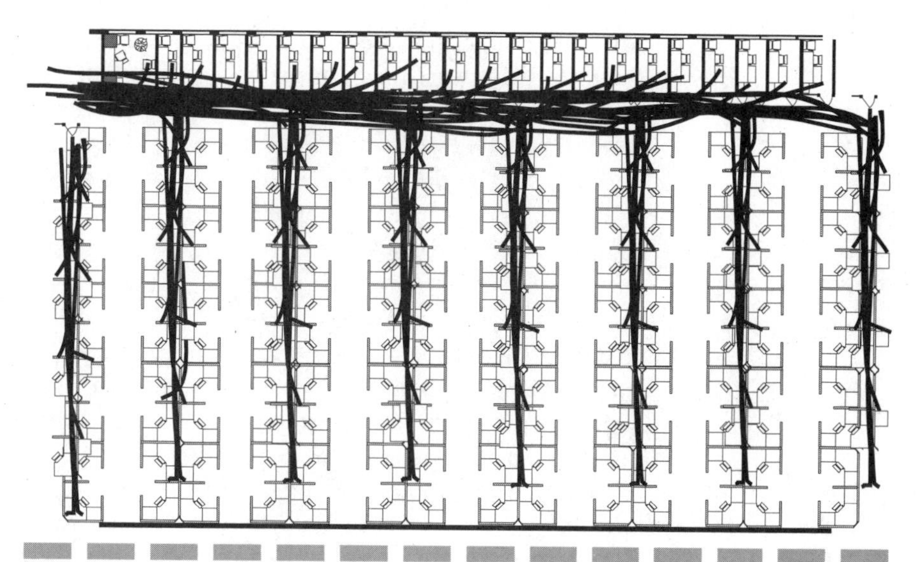

Figure 5.23.
Offices and cubicles were rewired and refurbished.

because the rewiring preceded recognition of this standard. This is illustrated in Figure 5.23.

Employees were relocated as wings were rebuilt. Two years elapsed as new PC-based applications outshone the mainframe and minicomputer applications. In the early 1990s the company embarked on a client/server applications to automate billing, boilerplate contract generation, and sales force automation. As a consequence, many day-to-day operations were moved to PCs and the network load increased further. While the client/server applications proved to be effective workflow solutions, the further success of e-mail, networked information bases, increased professional automation, and the client/server applications themselves pushed traffic levels above the saturation point for Ethernet at 10 Mbps. Although the wiring closets contained 10Base-T hubs with bridges and routers for connectivity to other segments, traffic levels and latencies were too high. Vendor solutions exchanged bandwidth for latencies of minutes.

Several client/server applications failed, not because they didn't work (because they really did), but because they created such high levels of new traffic that they were insupportable even on their own networks. Some of the PowerBuilder applications would run for hours mostly due to dropped traffic, collisions, and responses that timed out. The Gupta SQL servers running under NetWare were also no match for the load imposed by the inefficient data requests, and these systems withered under the sustained CPU and I/O loads. Solutions with duplexed and switched hubs were tried, but this was only a Band-Aid, and it ultimately failed because the organization was so distributed that the client/server applications needed to be distributed throughout the organization. The network bandwidth did not address the application overloads at the server, and if anything it made them more obvious. Many projects were abandoned. The consequences were consistent not with improper tool or network selection, but with improper design, implementation, and integration with the organizational culture.

Backbone traffic grew even more, given that that was possible. The company tried FDDI on fiber as a bandwidth fix, but the lack of a uniform infrastructure and a solution for long latencies between segments only served to increase some opportunities and crowd others from the corporate network. The company installed a series of FDDI backbones, and considered FDDI to the desktop. Costs and implementation problems halted this direction. With another corporate restructuring, the company changed direction. Management looked for outside advice.

This corporation is currently rewiring the wings again, this time with Category 5 wiring. Management foresees a need for ATM, frame relay, and ISDN to its 2000 remote locations. It foresees more integration with mail-enabled applications and more database-driven client/server applica-

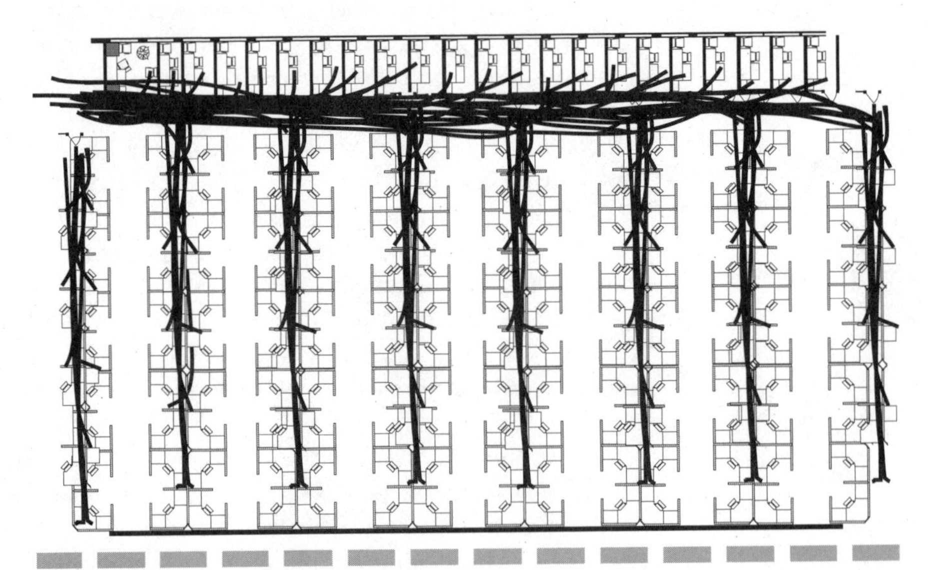

Figure 5.24.
Offices and cubicles were rewired and refurbished, this time for Category 5 support and fiber to each row of cubicles.

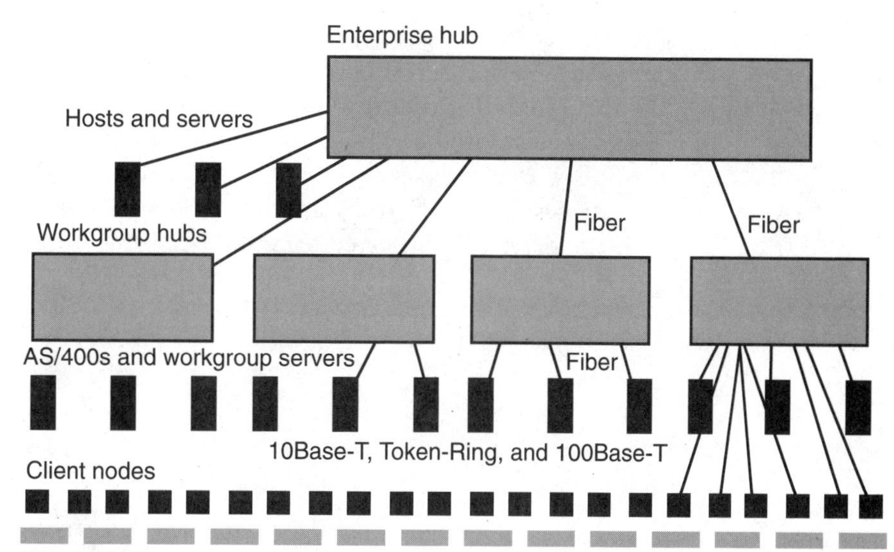

Figure 5.25.
The final logical architecture to support a new client/server initiative.

tions. As such, it is building a complete structured premises wiring system to support the bandwidth forecast for these new or reworked applications. This is illustrated in Figure 5.24.

If you notice that Figure 5.24 looks just like Figure 5.23, you are right. The need for additional pairs for 100Base-T support and future host or telecommunication and computer telephony integration forced a total replacement of the wiring. This aggravated management to no end, but they understand the concept of sunk costs very well. In fact, they realized that old conduits and old wiring left no room for the new. The old wires from two years before, even the 10Base5 backbones, were thus removed. Lighting is being replaced with more efficient ballasts and cooler high-intensity fixtures. Wings are being renovated. Perhaps the most important change in this process is that new client/server applications are being designed and optimized up front to run over modem lines, wireless, dedicated lines, frame relay, and ISDN, and not to require the massive bandwidths being installed at the headquarters. This will have a profound effect on the redeployment of new client/server applications because there will be bandwidth available for e-mail, file services, and normal office data processing too. The architecture that this corporation is working toward will look like the logical diagram in Figure 5.25.

This chapter explored some real-world examples of installing Fast Ethernet and migrating from slower local area network protocols, resolving bottlenecked designs, and dealing with the pitfalls of client/server applications. Fast Ethernet provides greater bandwidth for network traffic and addressed infrastructure limitations. These limitations can occur at intermediate nodes and other boundaries because of bandwidth limitations or excessive service delays. You need to understand the goals and strategies of the organization and match this within the limitations of the available Fast Ethernet technology.

GLOSSARY

100Base-FX The IEEE 802.3u physical layer device (PHY) sublayer using a two-strand fiber-optic standard for Ethernet signal transmission.

100Base-T The IEEE 802.3u extension for providing Ethernet transmission at 100 Mbps on twisted-pair and powered signal-regenerative hubs.

100Base-T4 The IEEE 802.3u physical layer device (PHY) sublayer using a four-pair wiring standard that can use voice-grade cable.

100Base-TX The IEEE 802.3u extension for providing duplexed Ethernet transmission at 100 Mbps on two pairs of twisted-pair wiring.

100BaseVG-AnyLAN An IEEE 802.12 committee extension for providing Ethernet transmission at 100 Mbps on twisted-pair, with quartet (MLT-5) signaling that uses a new MAC-layer protocol that does not support collision detection.

10Base-T A reference to the Ethernet IEEE 802.3 standard supplemental definition, specifically to twisted-pair wiring and connectors, twisted-pair variations, and signal-regenerative powered hubs. The number scheme designates that these networks are baseband networks with transmission rates of 10 Mbps. The maximum contiguous cable segment length is usually limited to 100 meters due to the extreme signal interference on the unshielded cabling. There are two versions. One supports bidirectional signaling with dual-pair telco wiring, thus allowing hardware to see collisions. The other version uses a single pair to support daisy-chaining of multiple workstations. Note that duplicate repeaters are required for the dual-pair telco.

10Base-TX The IEEE 802.3u physical layer device (PHY) sublayer using a two-pair wiring standard with UTP Category 5 or STP.

10Base2 A common reference to the Ethernet standard, specifically Cheapernet and Thinnet variations. The number scheme designates that these networks are baseband networks with transmission rates of 10 Mbps, with maximum contiguous coaxial segment lengths of 2×100 meters (actually 185 meters).

10Base5 An uncommon reference to the Ethernet standard. The number scheme designates that these networks are baseband networks with transmission rates of 10 Mbps, with maximum contiguous coaxial segment lengths of 5×100 (500 meters).

1Base5 The IEEE 802.3 reference to StarLAN.

4B/5B Encoding methods for packing four bits of data into every five bits transmitted over the data network. Encoding and decoding is performed by a standardized look-up table, which precludes certain recurring signal transitions. Fast Ethernet must achieve 125 MHz in signaling speed to provide a 100-MHz data rate.

5B/6B Encoding methods for packing five bits of data into every six bits transmitted over the data network. Encoding and decoding is performed by a standardized look-up table, which precludes certain recurring signal transitions. Fast Ethernet must achieve 120 MHz in signaling speed to provide a 100-MHz data rate.

6B/8B Encoding methods for packing six bits of data into every eight bits transmitted over the data network. Encoding and decoding is performed by a standardized look-up table, which precludes certain recurring signal transitions. Fast Ethernet must achieve 133.3 MHz in signaling speed to provide a 100-MHz data rate.

abnormal preamble A packet error that occurs when the preamble doesn't match the legal eight-byte Ethernet synchronization pattern.

ac Abbreviation for alternating current. Electricity.

active hub A multiported device that amplifies LAN transmission signals.

address 1. Data structure used to identify a unique entity, such as a particular process or network location. 2. A reference to a source or destination station on a network.

address error A packet improperly labeled with either source or destination information.

alignment error 1. In IEEE 802.3 networks, an error that occurs when a received frame's total number of bits is not divisible by eight. Alignment errors usually are caused by frame damage due to collisions. 2. A frame that has not been synchronized correctly.

American National Standards Institute The coordinating body for voluntary standards groups within the United States, this group is also a member of the International Organization for Standards. This is a governmental agency that maintains standards for science and commerce, including a list of acceptable standards for computer languages, character sets, connection compatibility, and many other aspects of the computer and data communications industries. Better known by the ANSI acronym.

American Wire Gauge A measurement system for electrical wire where larger numbers represent thinner wires (based upon the extrusion of a fixed amount of metal over longer distances). Abbreviated as AWG.

ANSI Acronym for American National Standards Institute (q.v.).

AnyNet IBM implementation of transport- and application-protocol independence on various networking environments and common traffic transport protocols. This protocol provides a method to build a uniform network transport environment, thereby bypassing some of the need for routers, encapsulation, and gateways. *See also multiprotocol transport networking.*

architecture The way hardware or software is structured, usually based on a specific design philosophy. Architecture affects both a computer's abilities and its limitations.

asynchronous communication Random transmission as needed.

Asynchronous Transfer Mode 1. A cell relay packet network providing from 25 Mbps to Gbits/s from central offices to central offices. 2. A form of packet switching; a subset of Cell Relay that uses 53-byte cells (five bytes of overhead, and another four bytes for LAN sequencing adaptation) as the basic transport unit. In concept, circuits of different signaling speeds can move data from desktop to desktop and across long-distance services without major changes in data format. Abbreviated as ATM.

ATM Acronym for Asynchronous Transfer Mode (q.v.).

attachment unit interface An IEEE 802.3 cable connecting the MAU (Media Access Unit) to the networked device. The term AUI also can be used to refer to the host back-panel connector to which an AUI cable might attach. Also called transceiver cable or drop cable.

attenuation Loss of communication signal energy.

AUI Acronym for Attachment Unit Interface (q.v.).

AUI cable The attachment unit interface cable that connects a workstation to a transceiver or fan-out box. Often called a drop cable.

auto-negotiation The algorithm that allows two or more nodes on a transmission link to negotiate transmission services. This usually relates to discovering the highest possible transmission speed, as in 10/100Base-T. *See also parallel detection.*

AWG Acronym for American Wire Gauge (q.v.).

backbone 1. *See bus.* 2. A collapsed (that is, minimized) wiring concentrator.

bandwidth 1. The range (band) of frequencies that are transmitted on a channel. The difference between the highest and lowest frequencies is expressed in Hertz (Hz) or millions of Hertz (MHz). 2. The wire speed of the transmission channel.

baseband A transmission channel that carries a single communications channel, on which only one signal can transmit at a given time.

baseline The normalized traffic or performance level. An interesting although perhaps flawed metric for network and computer performance management.

baselining The process of generating a baseline or making comparisons to a baseline.

bend radius The minimum bend allowable for fiber or cable. Assume a 10×diameter minimum unless it is otherwise defined. Minimum bend radius for PVC UTP is four inches. The minimum bend radius for Teflon is 10 inches.

bit An abbreviation for *b*inary dig*it*, A unit used in the binary numbering system; it can be 0 or 1.

bit error rate The percentage of erroneous transmitted bits received by a network node or server.

bits per second A rate at which data are transmitted over a communications channel. Abbreviated as bps.

blocking The delay that occurs when a channel is already physically switched and communications are established between end points thereby preventing other stations or nodes from sharing or interrupting access. This is primarily a function of network switching architectures, PBXs, and LAN switches. In a switching system, this is a condition in which no paths are available to complete a circuit. The term is also used to describe a situation in which one activity cannot begin until another has been completed.

bottleneck 1. Any obstruction that impedes completion of a task. 2. The critical path in task completion; the path without slack time. 3. The task or process in a larger system that has no slack time or is the task that delays subsequent and sequential tasks, so that system completion is delayed.

bps Acronym for bits per second.

break A physical break (electrical or optical) in the network media that prohibits passage of the transmission signal. *See also open.*

breakout bundle A single package containing multiple optical fibers, wires, and cables supporting the extraction of single wires at intermediate places along the full length of the wiring package.

bridge 1. A device that interconnects networks using similar protocols. A bridge provides service at level 2 of the OSI reference model. 2. An internetworking device that connects two networks at the media access control layer of the OSI reference model. Because the bridge does not read the network address of the packet, bridges are protocol-independent and (unlike routers or gateways) can route packets without understanding them. *See also gateway and router.*

broadcast storm An enterprise network event in which many broadcasts are sent at once, overloading intermediate nodes and creating a time-out or network panic. This is an undesirable network event in which many broadcasts are sent all at once, using substantial network bandwidth and typically causing network saturation at intermediate nodes. *See also cascade failure, network cascade, or network panic.*

burst A directed transmission from one network device to another that represents a significant portion of network transmission bandwidth.

bus A network topology in which nodes are connected to a linear configuration of cable.

cable A transmission medium of wires or optical fibers wrapped in a protective cover.

cable scanner A testing tool that verifies the integrity and performance of network wiring and cable. It tests for electrical breaks, shorts, impedance, capacitance, inductance, signal crosstalk, and signal attenuation. These tools are sometimes called pair or ring scanners when designed for Token-Ring or FDDI.

cable tester A testing tool that verifies the integrity and performance of network wiring and cable. It tests for electrical breaks, shorts, impedance, capacitance, as well as for signal crosstalk and signal attenuation. These tools are sometimes called pair or ring scanners when designed for Token-Ring or FDDI.

capacitance One of the electrical properties of the network cable and hardware.

Carrier Sense Multiple Access with Collision Detection 1. A communications protocol in which nodes contend for a shared communications channel and all nodes have equal access to the network. Simultaneous transmissions from two or more nodes result in random restarts of those transmissions. 2. A channel access mechanism wherein devices wishing to transmit first check the channel for a carrier. If no carrier is sensed for some period of time, devices can transmit. If two devices transmit at once, a collision occurs and is detected by all colliding devices, which subsequently delays their retransmissions for some random length of time. This access scheme is used by Ethernet and IEEE 802.3. 3. The Ethernet protocol. Abbreviated as CSMA/CD.

cascade The interconnection of wiring hubs or switches into other wiring hubs or switches to support and increase the network device count. Typically, such network support devices can be cascaded from two to five levels to conform within Ethernet and Fast Ethernet specifications. *See also cascade failure or network panic.*

cascade failure A sequential chain reaction overload of connectivity devices, or overt collapse of network channels and devices that cause shutdown of the enterprise network infrastructure. *See also network panic.*

Category 1 The TIA/EIA recommendation for two-pair twisted (TP) to support data transmission rates to 1 Mbps. This is voice-grade wire suitable for analog telephone, facsimile, and modem connections. It is not typically used for digital lines. It is not in any way similar to IBM Type 1 cable.

Category 2 The TIA/EIA recommendation for two-pair twisted-pair (TP) to support data transmission rates to 4 Mbps. This grade of wire is suitable for ARCNET and 4-Mbps Token-Ring.

Category 3 The TIA/EIA recommendation for two-pair twisted-pair (TP) to support data transmission rates to 10 Mbps. This is suitable for Ethernet and 4-Mbps Token-Ring.

Category 4 The TIA/EIA recommendation for two-pair twisted-pair (TP) to support data transmission rates to 16 Mbps. This is suitable for Ethernet and 16-Mbps Token-Ring, but not necessarily suitable for switched or full-duplex data transmission because of the near-end crosstalk and impedance problems. IBM Type 1 cable does correspond to these performance recommendations.

Category 5 The TIA/EIA recommendation for two-pair twisted-pair (TP) to support data transmission rates to 155 Mbps. This is suitable for TCNS, ARCNET PLUS, Fast Ethernet, FDDI, and ATM. IBM Type 1 (coaxial) cable may correspond to these performance recommendations.

CCITT Acronym for Consultative Committee for International Telephone and Telegraph (q.v.). *See ITM.*

channel 1. An individual communication path that carries signals. 2. The path between a sender and receiver (two PCs, for example) that carries one stream of data. The term also is used to describe the specific path between large computers and attached peripherals.

channel acquisition The process of acquisition, up to the point of establishing digital synchronization.

channel capacity 1. The amount of data that may be transferred across a channel stream per unit time. 2. the number of users that may be supported by a channel stream.

chatter The condition resulting when NIC electronics fail to shut down after a transmission and the NIC floods the network with random signals.

client 1. A network user or process, often a device or workstation. 2. A secondary processor that relies upon a primary processor or server.

coax *See coaxial cable.*

coaxial cable An insulated, tinned copper conducting wire surrounded by foamed PVC or Teflon, and shielded by tinned copper braid or an aluminum sleeve. It can carry data transmissions at very high data rates with little loss of information, and with a high immunity to outside interference.

collision 1. The event occurring when two or more nodes contend for the network at the same time. This is usually caused by the time delay that the signal requires to travel the length of the network. 2. The Ethernet protocol mechanism by which simultaneous or overlapping network access requests are handled. This network event is not an error in and of itself; it is an indication that two or more nodes attempted to transmit within the same slot. Unless the number of collisions is excessive, this is a normal condition.

collision detection A node's ability to detect when two or more nodes are transmitting simultaneously on a shared network.

collision enforcement The transmission of extra jam bits, after a collision is detected, to insure that all other transmitting nodes detect the collision.

congestion A slowdown in a network due to a bottleneck. Excessive network traffic. *See also traffic congestion.*

congestion error 1. An indication that a station does not have sufficient buffer space to copy a cell, frame, or packet addressed to it. 2. A packet sent to the file server, which then cannot buffer the frame, indicates that the server NIC may be considered congested. The board may require reconfiguration to allocate an additional incoming packet buffer.

Consultative Committee for International Telephone and Telegraph An international organization that makes recommendations for networking standards like X.25, X.400, and facsimile data compression standards. Abbreviated as CCITT. Now called the International Telecommunications Union Telecommunication Standardization Sector, and abbreviated as ITU or ITU-TSS.

contention An access method in which network devices compete for the right to access the physical medium. *See also token passing and circuit switching.*

controller The device in a network node that physically connects that device to the network media.

CRC Acronym for Cyclic Redundancy Check (q.v.).

CRC error Acronym for Cyclic Redundancy Check Error (q.v.).

CRC errors/s 1. An indication of a corrupted packet, either due to a faulty network interface board or a faulty cabling system. 2. A high number of CRC errors per second attributed to a single station indicates a faulty network interface board. If CRC errors are attributed to numerous stations on the network, this indicates a cabling problem.

crosstalk A technical term indicating that stray signals from other wavelengths, channels, communication pathways, or twisted-pair wiring have polluted the signal. It is particularly prevalent in twisted-pair networks or when telephone and network communications share copper-based wiring bundles. A symptom of interference that is caused by two cell sites causing competing signals to be received by the mobile subscriber. This can also be generated by two mobiles causing competing signals that are received by the cellular base station. Crosstalk sounds like two conversations, and often a distortion of one or the other or both.

CSMA/CD Acronym for Carrier Sense Multiple Access with Collision Detection (q.v.).

cut-through switch mode A switching technology that has the least latency because the MAC address is stripped from the header on-the-fly and the packet is redirected to the destination at wire speed. It provides no store-and-forward buffering. At high traffic loading or high error rates, the switch forwards bad packets, and drops all those it cannot immediately forward. *See also modified cut-through switch mode and store-and-forward switch mode.*

cycles per second *See Hertz.*

cyclic redundancy check A checksum—an error-checking algorithm that the transmitting station includes within a frame. The receiving station generates its own CRC to check against the transmitted CRC. If the results are different, the receiver usually requests a retransmission of the frame. This encoded value is appended to each frame by the data link layer to allow the receiving NIC to detect transmission errors in the physical channel. *See also frame check sequence.*

data When used in the context of communications, data refers to transmitted information, particularly information that is not interpreted by a particular protocol entity but merely delivered to a higher-level entity, possibly after some processing.

data rate The effective speed at which data is transferred over a transmission link. The data rate is the actual transmission rate after errors, synchronization, and overhead are factored out.

data-grade twisted-pair Telephone-type wire twisted over its length to preserve signal strength and minimize crosstalk and electromagnetic interference for high-speed data communications networks. Abbreviated as DTP or as data-grade TP.

dc Abbreviation for direct current. Electricity.

deference The process by which an Ethernet LAN delays its transmission when the channel is busy, to avoid contention with an ongoing network transmission. 100BaseT and 100BaseT4 use collision and signal detection, whereas 100BaseVG uses a round-robin access method.

delay The time between the initiation of a transaction by a sender and the first response received by the sender. Also, the time required to move a packet from source to destination over a given path.

demand priority An access method or the permission to transmit control algorithm for hub-based Ethernet network variants, including 10BaseVG-AnyLAN and some wireless protocols.

destination address The receiving station's address. *See address.*

device Any item on the network; this includes logical addresses that refer to software or hardware processes. It refers to any physical station, including a PC, workstation, mainframe, minicomputer, bridge, router, gateway, remote probe, repeater, or protocol analyzer.

Differential Manchester Encoding Digital coding scheme where a mid-bit-time transition is used for clocking, and a transition at the beginning of each bit time denotes a zero. The coding scheme used by 100BaseVG-AnyLAN.

drop cable Generally an 802.3 Ethernet transceiver cable. It may also refer to the Token-Ring lobe cable, or the cable from a hub to station on FDDI. *See transceiver cable and AIU cable.*

duplexed Ethernet A method to provide paired, full-duplexed Ethernet transmission circuits.

dynamic switching A switch that switches communicating nodes to a private circuit during the communication session. In effect, this provides LAN microsegmentation by connecting paired network devices to different subnets for the duration of a single message packet. It does not increase the network bandwidth per se, but can provide increased capacity through network partitioning and the application of dedicated bandwidth. *See also matrix switching, port switching, and segment switching.*

E-mail Electronic mail; a computer network and software by which messages can be sent to others on the network.

EIA Acronym for Electronic Industries Association (q.v.). *See Electronics Industry Association/Telecommunication Industry Association.*

EIA/TIA Acronym for Electronics Industry Association/Telecommunication Industry Association (q.v.).

EIA/TIA 568/569 Premise wiring recommendations for telecommunications and data communications in commercial and high-rise buildings and campus facilities. These are not standards or specifications, but represent rather a concept. Conformance testing to wiring category definitions is based on an industry-based general adaptation and consensus to these wiring concepts.

electromagnetic interference Interference by electromagnetic signals that can cause reduced data integrity and increased error rates on transmission channels. Electromagnetic interference is signal noise pollution from radio, radar, fluorescent lights, or electronic instruments. Abbreviated as EMI.

Electronics Industry Association/Telecommunication Industry Association Two industry groups merging telephonics and communications experience that have created a series of joint standards for consolidating premise wiring and measuring quality and performance. EIA was established in 1924 by common-carrier radio interests and the electronic industry community to respond to governmental regulatory matters. Abbreviated as EIA/TIA.

EMI Acronym for Electromagnetic Interference (q.v.).

encapsulation The wrapping of data in a particular protocol header. For example, Ethernet data is wrapped in a specific Ethernet header before network transmission. Also, a method of bridging dissimilar networks where the entire frame from one network is simply enclosed in the header used by the link-layer protocol of the other network.

Enet *See Ethernet.*

enterprise network A campus or wide area network that services all (or most) organizational sites supporting multiple point-to-point routes and/or integrating voice, facsimile, data, and video into the same channel. A (usually large and diverse) network connecting most major points in a company. An enterprise network differs from a WAN in that it is typically private and contained within a single organization.

error rate The number of errors during a period of time divided by the number of frames transmitted during that same interval. Errors may be reflected as a single bit error, but the rate should reflect errors as a basis not of bit throughput, but rather of frame throughput. *See also throughput.*

Ether The lumeniferous ether was the omnipresent passive medium theorized in 1765 by Christiaan Huygens. It was proposed as the medium that carried light (electromagnetic) waves from the Sun to the Earth.

Ethernet A popular baseband local area network from which the IEEE 802.3 standard was derived. Ethernet applies the IEEE 802.2 MAC protocols and uses the persistent CSMA/CD protocol. It is built on a bus topology based on original specifications invented by Xerox Corporation and developed jointly by Xerox, Intel, and Digital Equipment Corporation. Sometimes abbreviated as Enet.

Ethernet address A coded value indicating the manufacturer, machine type, and a unique identifying number. There are source and destination addresses in every valid Ethernet packet.

Ethernet controller An interface device that provides protocol access for computer equipment to a network. Each node on the network must have an Ethernet controller.

European Computer Manufacturers Association A group of European computer vendors that have done substantial OSI standardization work. Abbreviated as ECMA.

Fast Ethernet A reference to Ethernet running at 100 Mbps and based on either the 100Base-T or 100BaseVG-AnyLAN IEEE standards.

fast switching A feature whereby a route cache is used to expedite packet switching through a router.

fast-switched hub A network device that retransmits a frame from one network segment or port to another segment or port. By checking the destination address field in the frame header before transmission, the hub can redirect the frame to the segment or port where the destination device is located.

FCC Acronym for Federal Communications Commission.

FDDI *See Fiber Data Distributed Interchange (q.v.).*

FDDI II The proposed NSI standard to enhance FDDI with isochronous transmission for connectionless data circuits and connection-oriented voice and video circuits.

feature detector A device for measuring lengths, objects, scatter, and probable performance in optical fiber.

FEP Acronym for Fluorinated Ethylene Propylene (q.v.).

fiber Optical fiber; a network signaling medium in which plastic or glass fiber transports optical signals.

Fiber Data Distributed Interchange An optical fiber network based on the ANSI X3.139, X3.148, X3.166, X3.184, X3.186, or X3T9.5 specifications. FDDI provides a 125 Mbps signal rate with four bits encoded into five-bit format for a 100-Mbps transmission rate. It functions on single-ring, dual-ring, and star networks with a maximum circumference of 250 km. Copper-based hardware is also an option. Abbreviated as FDDI. *See also CDDI, SDDI, TP-DDI, and TP PMD.*

Fiber Optical Inter-Repeater Link The interconnection protocol required for point-to-point repeater links based on optical fiber in LANs. Abbreviated as FOIRL.

fiber optics Thin glass or plastic cables that transmit data in the form of pulse-modulated light beams.

fiber-optic cable A thin, flexible medium capable of conducting modulated light transmission. Compared with other transmission media, fiber-optic cable is more expensive, impervious to electromagnetic interference, and capable of higher data rates.

Fibre Channel A high-end networking medium and protocol based on parallel and synchronized transmission of light through optical fibers for point-to-point connectivity.

fluorinated ethylene propylene Better known as Teflon. This high-temperature insulator is used in cable coatings and foam compositions where building codes specify fire-resistant, high-temperature applications. Abbreviated as FEP.

FOIRL Acronym for Fiber Optic Inter-Repeater Link (q.v.).

forced collision A collision that occurs when a packet is transmitted even if traffic (carrier sense) is detected on the network; that is, if the packet will collide with other packets already on the network. When a packet is transmitted and collides, it is received at the destination node with either a CRC error or an alignment error, if it is received at all.

fragment 1. A partial cell, frame, or packet. 2. A normal Ethernet event due to a collision. 3. A sudden increase in Ethernet fragments can indicate a problem with a network component. A good rule of thumb is to examine fragments in conjunction with utilization. If both Ethernet utilization and fragments increase, this indicates an increase in usage of the cabling system. If fragments increase and utilization remains steady, however, this indicates a faulty network component.

fragmentation The process of breaking a packet into smaller units when transmitting over a network medium that cannot support the original size of the packet.

frame 1. A self-contained group of bits representing data and control information. The control information usually includes source and destination addressing, sequencing, flow control, preamble, delay, and error control information at different protocol levels. 2. A frame may be a packet with framing bits for preamble and delay. 3. Data transmission units for FDDI. The terms packet, datagram, segment, and message are also used to describe logical information groupings at various layers of the OSI reference model and in various technology circles.

frame check sequence 1. The encoded value appended to each frame by the data link layer to allow receiving Ethernet controllers to detect transmission errors in the physical channel. Also called a Cyclic Redundancy Check. 2. An HDLC term adopted by subsequent link-layer protocols and referring to extra characters added to a frame for error-control purposes. Abbreviated as FCS.

framing The process of assigning data bits into the network time slot.

frequency Measured in Hertz (Hz), the number of cycles of an alternating current signal per unit time.

ft. Abbreviation for an SAE foot.

functional cascade The unanticipated side effects resulting from event- or trigger-driven code that begins a sequence of executing unrelated code or iteratively executing the same functions because object values in OOP code have been changed by prior functions.

gateway 1. A device that routes information from one network to another. It often provides an interface between dissimilar networks and provides protocol translation between the networks, such as SNA and TCP/IP or IPX/SPX. A gateway is also a software connection between different networks; this meaning is not implied in this book. The gateway provides service at levels 1 through 7 of the OSI reference model. 2. In the IP community, an older term referring to a routing device. Today, the term router is used to describe nodes that perform this function, and gateway refers to a special-purpose device that performs a Layer 7 conversion of information from one protocol stack to another. *See also Bridge and Router.*

Gbps Abbreviation for gigabits (1,000,000,000 bits) per second.

gridlock The adverse performance condition that occurs when traffic from overlapping or intersection channels prevents the free flow of traffic on the other channel(s).

hop 1. The routing of a cell, frame, or packet through a network device and/or transmission channel based on destination address information. 2. The passage of a packet through one bridge, router, switch, or gateway.

hub 1. A network interface that provides star connectivity. 2. A wiring concentrator. 3. Generally, a term used to describe a device that serves as the center of a star-topology network. In Ethernet/IEEE 802.3 terminology, a hub is an Ethernet multiport repeater, which is sometimes referred to as a concentrator. The term is also used to refer to a hardware/software device that contains multiple independent but connected modules of network and internetwork equipment. Hubs can be active (where they repeat signals sent through them) or passive (where they do not repeat, but merely split, signals sent through them). *See MAU.*

hub adapter A network interface board that provides access for additional network nodes. This network interface unit essentially doubles as a wiring concentrator inside another device, such as a PC.

I/O Abbreviation for input and output from a computer. Refers to all memory movement on the bus, data moved to and from disks, screen and speaker output, and data frames transmitted to and from the network channel.

I/O channel *See bus.*

IBM cabling system A system for specifying different types of wire; most wire in IBM-wired systems is Type 1 or Type 3. Not to be confused with the TIA/EIA category recommendation system.

IEEE Acronym for the Institute for Electrical and Electronic Engineers (q.v.).

IEEE 802 An Institute for Electrical Engineering standard for interconnection of local area networking computer equipment. The IEEE 802 standard describes the physical and data link layers of the OSI reference model.

IEEE 802.1 A specification for media-layer physical linkages and bridging.

IEEE 802.1d The standard that detects and manages logical loops in a network. When multiple paths exist, the bridge or router selects the most efficient one. When a path fails, STA automatically reconfigures the network with a new active path. Also known as the Spanning Tree Algorithm.

IEEE 802.11 A physical- and MAC-layer specification for wireless network transmission at transmission speeds from 1 to 4 Mbps. This specification includes the basic rate set for fixed bandwidths supported by all wireless stations (for compatibility) and an extended rate set for optional speeds. Another addendum includes a dynamic data rate set.

IEEE 802.12 A 100-Mbps Ethernet specification based on four wire pairs and quartet signaling.

IEEE 802.2 A specification for media-layer communication typified by Ethernet, FDDI, and Token-Ring. *See also logical link control.*

IEEE 802.3 An Ethernet specification derived from the original Xerox Ethernet specifications. It describes the CSMA/CD protocol on a bus topology using baseband transmissions.

IEEE 802.3u The Fast Ethernet specification for 100-Mbps transmission on two pairs, four pairs, or fiber.

IEEE 802.5 A token ring specification derived from the original IBM Token-Ring LAN specifications. It describes the token protocol on a star/ring topology using baseband transmissions.

IEEE 802.9a A recommendation for isochronous Ethernet with a standard 10-Mbps Ethernet channel and a second 6-Mbps isochronous channel for simultaneous videoconferencing using the BRI ISDN interface. This recommendation also provides for connectivity to WAN channels using ISDN, Switched 56, or T-1 services with a CSU/DSU interface.

impedance The mathematical combination of resistance and capacitance that is used as a measurement to describe the electrical properties of the coaxial cable and network hardware.

inductance The property of electrical fields to induce a voltage to flow on the coaxial cable and network hardware. It is usually a disruptive signal that interferes with normal network transmissions.

infrastructure 1. The premise wiring plant supporting data communications; the risers, jumpers, patch panels, and wiring closets; the hubs, PBXs, and switches; the computer operations providing services to network devices; the network division; the test equipment; and the design and support staff maintaining operations. 2. The physical and logical components of a network. Typically, this includes wiring, wiring connections, attachment devices, network nodes and stations, interconnectivity devices (such as hubs, routers, gateways, and switches), operating environment software, and software applications.

Institute for Electrical and Electronic Engineers A membership-based organization based in New York City that creates and publishes technical specifications and scientific publications. Abbreviated as IEEE.

intelligent hub A wiring concentrator for Ethernet, FDDI, and Token-Ring nodes that isolates failing lobes, trunks, and nodes and sometimes gathers performance and error statistics for storage in an MIB in local RAM.

interconnection A junction (telecommunication) connecting two communication carriers (such as between cellular and land-line networks) allowing mutual access by customers to each carrier's network.

interconnectivity The process where different network protocols, hardware, and host mainframe systems can attach to each other for transferring data to each other.

interface A device that connects equipment of different types for mutual access. Generally, this refers to computer software and hardware that enable disks and other storage devices to communicate with a computer. In networking, an interface translates different protocols so that different types of computers can communicate together. In the OSI model, the interface is the method of passing data between layers on one device.

interference Unwanted communication channel noise.

interframe spacing The 96-bit waiting time between transmissions to allow receiving Ethernet controllers to recover. This corresponds to

96 ms at 10 Mbps and 9.6 ms at 100 Mbps. PACE and priority demand systems support the interface spacing only for interoperability with standard Ethernet.

intermediate node Any network device that is not a primary server, supplemental server, or end-user workstation, that provides packet replication, encapsulation, translation, or redirection. Examples include repeaters, bridges, routers, switches, and gateways. Abbreviated as IN.

International Standards Organization The standards-making body responsible for OSI, a set of communications standards aimed at global interoperability. The United States is one of 75 member countries. Abbreviated as ISO.

International Telecommunications Union Previously named the CCITT (Comite Consultatif International Telegraphique et Telephonique). This is a leading group to develop telecommunications standards. Abbreviated as ITU or ITU-TTS.

International Telecommunications Union-Telecommunications Standards Sector Formerly, the Consultative Committee for International Telephone and Telegraph (CCITT), this international organization makes recommendations for networking standards like X.25, X.400, and facsimile data compression standards. Abbreviated as ITU-TSS.

internet address 1. An address in the following format: username@system.type. 2. An address applied at the TCP/IP protocol layer to differentiate network stations from each other. This is in addition to the station hardware or protocol address. 3. Also called an "IP address," at least a 32-bit address (could be 128-bit) assigned to hosts using TCP/IP. The address is written as four octets separated with periods (dotted decimal format) that are made up of a network section, an optional subnet section, and a host section.

internetwork A collection of networks interconnected by routers that functions (generally) as a single network. Sometimes called an internet, which is not to be confused with the Internet.

internetworking A general term used to refer to the industry that has arisen around the problem of connecting networks with each other. The term can refer to products, procedures, and technologies.

interoperability The process where different network protocols, network hardware, and host mainframe systems can process data together.

interpacket delay The time between arrivals of packets on the network.

ISO Acronym for the International Standards Organization (q.v.).

isochronous communication The consistent and uninterrupted stream of data across the network that is the necessary form of communication for video sessions.

isochronous Ethernet 10-Mbps Ethernet with an ISDN basic rate interface for video sessions. *See also IEEE 802.9a.*

isolation switch A remote (out-of-band) radio-frequency trigger relay that electrically switches sections of network wiring or coaxial cable and alters network topology. *See also isolation switch or firewall.*

ITU Acronym for International Telecommunications Union (q.v.).

ITU-TSS Acronym for International Telecommunications Union-Telecommunications Standards Sector (q.v.).

jabber 1. To talk without making sense. 2. The condition that results when a transceiver's carrier-sense electronics malfunction and the transceiver broadcasts in excess of the specified time limit, thus creating an oversized frame.

jabber frame A frame that exceeds 1518 bytes in the data field and violates the IEEE 802.3 specifications.

jam A short encoded sequence emitted by the transmitting node to ensure that all other nodes have detected a collision, and used for collision enforcement.

jitter A network failure that occurs when network segment preamble and the frame signal are out of phase. The jitter shows up as transmission signal distortion, decay, frequency errors, and timing errors on all DC-based hubs or switches. Analog signal distortion is caused by variation of the signal from the timing positions. Jitter can cause data loss, particularly at high speeds.

jumper A user-supplied cable that connects a workstation to a wallplate or a punchdown panel to a patch panel. Also called a patch cord.

KB Acronym for kilobytes (1024 bytes) of memory. Also K.

Kbps Abbreviation for kilobits (1000 bits) per second.

kHz *See kiloHertz.*

kiloHertz A measure of audio and radio frequency (a thousand cycles per second). Abbreviated as kHz. The human ear can hear frequencies up to about 20 kHz. There are 1000 kHz in 1 MHz.

kilometer A unit representing 1000 m, or approximately 3200 SAE feet. It also may be represented by km.

km Acronym for kilometer (q.v.).

LAN Acronym for Local Area Network (q.v.).

LAN switch The hublike intermediate-node hardware used for increasing overall network performance either through static microsegmentation and dynamic port switching. Typically, network devices are dynamically paired through the LAN switch for each packet transmission only for the duration of that single transmission. This technology is typically applied for LAN-to-ATM connectivity. *See also dynamic switching, matrix switching, and segment switching.*

late collision A collision indicated by an oversized runt frame, usually indicative of a network exceeding length or size specifications.

latency 1. The waiting time for a station desiring to transmit on the network. 2. The delay or process time that prevents completion of a task. Latency usually refers to the lag between request for delivery of data over the network until it is actually received. 3. The period of time after a request has been made for service before it is fulfilled. 4. The amount of time between when a device requests access to a network and when it is granted permission to transmit.

Lattisnet The first product from Synoptics in 1985, and the basis for the development of the 10Base-T standard. This early media variant of 802.3 Ethernet used unshielded twisted-pair and powered signal-regenerative hubs, proving that networking protocols could be media-independent.

linkage product 1. Any unit that provides an interface between network segments or different types of operating systems and equipment. This includes gateways, bridges, routers, and other specialty components. 2. Any unit that provides either interconnectivity or interoperability. *See also intermediate node.*

load balancing 1. A technique to equalize the workload over peer and client network components. This includes workstations, storage disks, servers, network connectivity devices (such as bridges, routers, gateways, and switches), and network transmission channels. 2. In routing, the ability of a router to distribute traffic over all its network ports that are the same distance from the destination address. Good algorithms use both line speed and reliability information. Load balancing increases the utilization of network segments, thus increasing effective network bandwidth.

lobe A section of cable or wire extending from an MAU, hub, or concentrator to a network station.

local area network A network limited in size to less than a city block. This usually services from 2 to 100 users. Abbreviated as LAN.

logical channel A nondedicated, packet-switched communications path between two or more network nodes. Through packet switching, many logical channels can exist simultaneously on a single physical channel.

logical device 1. Any addressable node on a network. 2. A description that lists how the network references physical devices.

long frame A frame that exceeds the specified protocol length maximum.

long packet A packet that exceeds maximum packet size including address, length, and CRC fields.

loopback test A test for faults over a transmission medium where received data is returned to the sending point (thus traveling a loop) and compared with the data sent.

m Acronym for meter (q.v.).

MAN Acronym for metropolitan area network (q.v.).

Manchester Encoding A digital encoding technique in which there is a transition in the middle of each bit time period. A "1" is represented by a high level during the first half of the bit time period, whereas a "0" is represented by a low level during the first half of the bit time period. The coding scheme used by IEEE 802.3 Ethernet at 10 Mbps. 4B/5B or 5B/6B with NRZ are used for Fast Ethernet variants.

matrix switch A cross-point switch that establishes logical connections for communications. *See also virtualization and switched virtual connection.*

MAU Acronym for Media Access Unit (q.v.). or Multi-Station Access Unit (q.v.).

MB Abbreviation for megabyte (1024 kilobytes) of data.

Mbps Acronym for megabits (1,000,000 bits) per second. Not a binary measurement.

media 1. The physical material used to transmit the network transmission signal. For wired networks, it is some form of copper wire or optical fiber. However, wireless networks use infrared or radio-frequency signals with the air as the medium. 2. The substance used by a physical data-storage device to record data; magnetic tape, floptical disks, or CD-ROM, for example.

Media Access Unit A device that connects directly to a lobe wire, broadcasts and receives information over that cable, and switches the signals to the next active downstream station. It is abbreviated as MAU. *See also hub.*

megabits per second The number of millions of bits tranferred per second. Abbreviated as Mbps.

MegaHertz Signal frequency use for voice, data, TV, and other forms of electronic communications in the millions of cycles per second. Abbreviated as MHz.

message 1. Any cell, frame, or packet containing a response to a LAN-type network request, process, activity, or network management operation. 2. A PDU of any defined format and purpose. 3. An application-layer logical grouping of information. *See also datagram, packet, frame, payload, and segment.*

meter A unit of measurement equivalent to 39.25 SAE inches, or 3.27 ft. Meter is abbreviated m.

metric A formal measuring standard or benchmark. Network performance metrics include Mbps, throughput, error rates, and other less formal definitions.

metropolitan area network A network that spans buildings, or city blocks, or a college or corporate campus. Optical fiber repeaters, bridges, routers, packet switches, and PBX services usually supply the network links. Abbreviated as MAN.

MHz Acronym for MegaHertz (q.v.).

microsecond 1×10^{-6} second. Abbreviated as µs.

microsegmentation 1. The division of a large network into smaller isolated (and bridged) segments. 2. A typical method to decrease network performance saturation on LANs that have become a traffic bottleneck. 3. The process of increasing bandwidth and managing loads by segmenting LANs into smaller units, establishing a firewall between these smaller units, and routing or Switching traffic as necessary between units.

millisecond 1×10^{-3} second, abbreviated ms.

misaligned frame A frame that trails a fragmentary byte (1-7 residual bits), and has an FCS error, or an Ethernet packet that was framed improperly by the receiving station and is therefore a synchronization error.

MLT-3 Acronym for Multi-Level Transmission-3 (q.v.).

MLT-5 Acronym for Multi-Level Transmission-5 (q.v.).

modified cut-through switch mode A switching technology that has a latency equivalent to that of bridges because it reads the entire packet header. It can retry a failed transmission more than once, because it buffers that single packet. Packets that arrive for the same destination during this interval are dropped, because they are not buffered. *See also cut-through switch mode and store-and-forward switch mode.*

modular wiring *See premise wiring or IEA/TIA 568/569.*

monitor *See Protocol Analyzer.*

µs Abbreviation for microsecond.

ms Acronym for millisecond (q.v.).

Multi-Level Transmission-3 A signal encoding method used on FDDI on twisted-pair that randomizes data to reduce electrical emissions, and also equalizes the signal levels. Abbreviated as MLT-3.

Multi-level Transmission-5 A signal encoding method used on 100-Mbps Ethernet on four-pair unshielded twisted-pair that randomizes data to reduce electrical emissions and also equalizes the signal levels. Abbreviated as MLT-5.

multicast 1. The ability to broadcast to a select subset of nodes. 2. Single packets copied to a specific subset of network addresses. These addresses are specified in the destination-address field. In contrast, in a broadcast, packets are sent to all devices in a network.

multimeter A test tool that measures electrical voltages (units in V) and resistances (units in W). It is also called multitester. It is sometimes called an ohmmeter.

nanosecond 1×10^{-9} second. Abbreviated as ns or ns.

NAU Acronym for network access unit (q.v.) or network addressable unit (q.v.).

near-end crosstalk Signal interefence on twisted-pair, usually due to unwinding the pairs too much for attachment to a block or connector. Abbreviated as NEXT.

network 1. Hardware and software that allows computers to transmit data over both local and long distances. 2. An area comprising two or more MSAs (Metropolitan Statistical Areas) and portions of any RSA (Rural Service Area) or RSAs located between such MSAs, in which the company provides or plans to provide uninterrupted cellular telephone service to system users traveling in and between such MSAs. The creation of a network typically leaves intact the ownership interests in the respective systems that comprise the network, but allows the systems to be combined for operational and marketing purposes. 3. A collection of computers and other devices that are able to communicate with each other over some network medium.

network access unit The network controller. Abbreviated as NAU. *See also controller.*

network address Also called a protocol address; a network layer address referring to a logical, rather than a physical, network device.

network administrator A person who helps maintain a network.

network analyzer A device offering various network troubleshooting features, including protocol-specific packet decodes, specific preprogrammed troubleshooting tests, packet filtering, and packet transmission. *See protocol analyzer.*

network cascade An enterprise network event in which many broadcasts are sent at once, overloading intermediate nodes and creating a time-out or network panic. This is an undesirable network event in which many broadcasts are sent all at once, using substantial network bandwidth and, typically, causing network saturation at intermediate nodes. *See also broadcast storm, cascade failure, or network panic.*

network interface card The network access unit that contains the hardware, software, and specialized PROM information necessary for a station to communicate across the network. Usually referenced as network interface controller. Abbreviated NIC.

network interface unit *See controller or network interface card.*

network monitor *See protocol analyzer.*

network operating system 1. A platform for networking services that combines operating system software with network access. This is typically not application software, but rather an integrated operating system. 2. The software required to control and connect stations into a functioning network conforming to protocol and providing a logical platform for sharing resources. Abbreviated as NOS.

network panic 1. A sequential chain-reaction overload of connectivity devices or overt collapse of network channels and devices that causes shutdown of the enterprise network infrastructure. 2. An enterprise network event in which many broadcasts are sent at once, overloading intermediate nodes and creating a time-out or network panic. This is an undesirable network event in which many broadcasts are sent all at once, using substantial network bandwidth and, typically, causing network saturation at intermediate nodes. *See also broadcast storm, cascade failure, or network cascade.*

network virtualization The creation of dedicated logical point-to-point connections between various devices on a network through the use of segment switching technology.

NEXT Acronym for near-end crosstalk (q.v.).

NIC Acronym for network interface card (q.v.).

NIU Acronym for network interface unit (q.v.).

no backoff error A transmission state that results if a transceiver transmits when there is no carrier, but does not wait for the necessary 96-bit delay.

node 1. A logical, nonphysical interconnection to the network that supports computer workstations or other types of physical devices on a network that participates in communication. 2. Alternatively, a node may connect to a fan-out unit providing network access for many devices. A device might be a terminal server or a shared peripheral such as a file server, printer, or plotter.

non-persistent A property of a MAC protocol by which a MAC entity shall sample the state of the medium at random intervals and shall only attempt to access the medium if the medium is in an idle state at the time of sampling.

NOS Acronym for network operating system.

NRZ Acronym for non return to zero.

NRZI Acronym for non return to zero, invert on ones (q.v.).

ns Acronym for nanosecond (q.v.).

ohmmeter *See multimeter.*

open A partial physical break (electrical or optical) through one or more signal conductors in the network media that prohibits passage of the transmission signal. Open circuits usually do not refer to a complete cut of the media. *See also break.*

open circuit A broken path along a transmission medium. Open circuits will usually prevent network communication.

Open Systems Interconnection reference model A specification definition from the International Standards Organization. It is a data communication architectural model for networking. Abbreviated as the OSI model.

optical fiber *See fiber.*

optical time-domain reflectometer Test equipment that verifies proper functioning of the physical components of the optical fiber network with a sequence of time-delayed optical pulses. Primarily, this tool checks the contiguity and signal-carrying capacity of optical fiber for data communications. *See also time-domain reflectometer.*

OS Acronym for operating system (q.v.).

OSI Acronym for Open Systems Interconnection reference model (q.v.).

OSPF Acronym for open shortest path first (q.v.).

OTDR Acronym for optical time-domain reflectometer (q.v.).

oversize error A frame or packet greater than the largest allowable size. Oversized packets indicate that a faulty or corrupted driver is in use on the network.

oversized frame A frame that exceeds the maximum frame size defined by a protocol.

oversized packet A packet that exceeds the maximum packet size including address, length, and CRC fields.

PACE Acronym for priority access control enabled (q.v.).

packet A self-contained group of bits representing data and control information. A logical grouping of information that includes a header and (usually) user data. The control information usually includes source and destination addressing, sequencing, flow control, and error control information at different protocol levels. This term generally refers to data transmission units for ARCNET, Ethernet, Token-Ring, and switching. *See also cell, datagram, frame, message, NPDU, payload, and segment.*

packet buffer A structure created in computer memory to build, disassemble, or temporarily store network data frames.

packet burst An overwhelming broadcast of frames requesting network and station status information, requesting source or destination addresses, or indicating panic error messages.

packet switching A network transmission methodology that uses data to define a start and length of a transmission for digital communications. A process of sending data in discrete blocks. A network on which nodes share bandwidth with each other by intermittently sending logical information units (packets). In contrast, a circuit-switching network dedicates one circuit at a time to data transmission. *See also circuit switching and message switching.*

packet-switching network Most commonly, packet-switched networks refer to X.25, ATM, or Frame Relay. Packet-switched networks offer flexibility for multipoint connections, high reliability, and flexible pricing.

pair scanner A testing tool that verifies the integrity and performance of network wiring and cable. It tests for electrical breaks, shorts, impedance, and capacitance, as well as for signal crosstalk and signal attenuation. These tools are sometimes called pair or ring scanners when designed for Token-Ring and FDDI.

parallel detection The ability to distinguish 100Base-TX from 100Base-T4 and the NLP from the FLP burst sequences.

patch cord A flexible wire unit or element with quick-connects used to establish connections on a patch panel.

patch panel A cross-connect device designed to accommodate the use of a patch cord for the simplification of wiring additions, movements, and other changes.

peak 1. A high volume of network traffic. 2. A local or absolute traffic volume maximum.

persistence A statistical term referring to a protocol's method of accessing the network. 10- and 100-Mbps Ethernet is persistent in transmitting, while 100BaseVG-AnyLAN with demand priority or PACE are nonpersistent and wait for a permission slot.

phantom voltage The voltage differential (5 volts dc) maintained between the transmit and receive wire pairs in a 100Base-T or 100BaseT4 twisted-pair network.

phase-locked loop signal A transmission method for timing signals used on 10Base-T and 100Base-T wiring hubs and NICs. Abbreviated as PLL.

physical address The unique address associated with each workstation on a network. A physical address is devised to be distinct from all other physical addresses on interconnected networks. A worldwide designation unique to each unit.

physical device Any item of hardware on the network.

PLL Acronym for phase-locked loop signal (q.v.).

polling An access method in which a primary network device inquires, in an orderly fashion, whether secondary nodes have data to transmit. The inquiry occurs in the form of messages to each secondary that gives the secondary the right to transmit.

polyvinyl chloride. An extensively used insulator in cable coatings and coaxial cable foam compositions. PVC insulation does not meet fire code for installation in plenums in some jurisdictions. Abbreviated as PVC. *See Teflon.*

port switching A switch that switches communicating nodes to a private circuit during the communication session. In effect, this provides LAN microsegmentation by connecting paired network devices to different subnets for the duration of a single message packet. It does not increase the network bandwidth per se, but can provide increased capacity through network partitioning and the application of dedicated bandwidth. *See also dynamic switching, matrix switching, and segment switching.*

position-dependent unfairness A situation common to networks where some stations receive better service due to proximity to other stations or a central location in a bus-structured LAN, a hub, or packet switch.

preamble The 64-bit encoded sequence that the physical layer transmits before each frame to synchronize clocks and other physical layer circuitry at other nodes on the Ethernet channel.

premise wiring A telecommunications and data communications wiring infrastructure that embodies the concept of flexible, recyclable, reconfigurable, and reusuable modular components centered in the wiring closet.

Priority Access Control Enabled A 3Com modification to the Ethernet network access protocol to disable the collision detection mechanism and replace it with a hub-based bandwidth allocation. This MAC-layer protocol implements bandwidth allocation for time-sensitive multimedia delivery. Abbreviated as PACE.

propagation delay The time required for data to travel over a network from its source to its ultimate destination.

protocol A formal set of rules by which computers can communicate including session initiation, transmission maintenance, and termination.

protocol address *See network address.*

protocol analyzer Test equipment that transmits, receives, and captures Ethernet packets to verify proper network operation.

PVC *See polyvinyl chloride.*

quartet signaling The transmission method 100BaseVG-AnyLAN applies whereby the signal is split over and recombined from multiple wires in order to lower the power and signal Hz.

queue A waiting line with requests for service. Generally, an ordered list of elements waiting to be processed. In routing, a backlog of packets waiting to be forwarded over a router interface.

queueing delay The amount of time that data must wait before it can be transmitted onto a statistically multiplexed physical circuit. *See also latency.*

radio frequency interference Electronically propagated noise from radar, radio, or electronic sources. *See electromagnetic interference.*

radio frequency switch A remote radio frequency trigger relay that electrically switches sections of network and alters network topology. *See also matrix switch.*

radio-frequency interference Electronically propagated noise from radar, radio, or electronic sources. Abbreviated as RFI. *See also electromagnetic interference.*

repeater 1. A device that boosts a signal from one network lobe or trunk and continues transmission to another similar network lobe or trunk. Protocols must match on both segments. The repeater provides service at level 1 of the OSI reference model. 2. A device that regenerates and propagates electrical signals between two network segments.

request/response unit *See RU reverse channel.*

resistance The measurement of the electrical properties of the wiring and network hardware that describes their ability to hinder passage of electrons.

RF Acronym for radio frequency (q.v.).

RFI Acronym for radio-frequency interference (q.v.).

RG-58 Coaxial cable with 50-ohm impedance. Used by IEEE 802.3 10Base2.

RG-62 Coaxial cable with 93-ohm impedance. Used by ARCNET.

ring 1. The connection and call signaling method on POTS. 2. A network topology that has stations in a circular configuration.

ring topology Topology in which the network consists of a series of repeaters connected to one another by unidirectional transmission links to form a single closed loop. Each station on the network connects to the network at a repeater.

RJ-11 Standard four-wire connectors for phone lines.

RJ-22 Standard four-wire connectors for phone lines with secondary phone functions (such as call forward, voice mail, or dual lines).

RJ-45 Standard eight-wire connectors for networks. Also used as phone lines in some cases. *See also ISO/IEC 8877.*

RMON Acronym for remote monitoring (q.v.).

route A path through an internetwork.

router 1. A device that interconnects networks that are either local area or wide area. 2. A device providing intercommunication with multiple protocols. 3. A device providing service at level 3 of the OSI reference model. 4. A router examines the network address of each packet. Those packets that contain a network address different from the originating PCs address are forwarded onto an adjoining network. Routers also have network-management and filtering capabilities, and many newer routers incorporate bridging capabilities as well. *See also Bridge and Gateway.*

routing The process of finding a path to the destination host. Routing is very complex in large networks because of the many potential intermediate destinations a packet might traverse before reaching its destination host.

runt frame An Ethernet frame that is too short. A runt frame has fewer than the 60 bytes in the data fields required by the IEEE 802.3. If the frame length is less than 53 bytes, a runt frame indicates a normal collision. A frame less than 60 bytes, but at least 53 bytes, indicates a late collision.

s Abbreviation for second.

saturation An occurrence on networks or network devices when packets arrive on the network faster than they can be handled.

scanner A testing tool that verifies the integrity and performance of network wiring and cable. It tests for electrical breaks, shorts, impedance, and capacitance, as well as for signal crosstalk and signal attenuation. These tools are sometimes called pair or ring scanners when designed for FDDI or Token-Ring.

segment switching A switch that provides LAN microsegmentation by connecting network devices to different subnets. It does not increase the bandwidth per se, but can provide increased capacity through network partitioning. *See also dynamic switching, matrix switching, and port switching.*

segmentation The process of fragmenting an SN-Data PDU or SN-Unitdata PDU into a number of LSDUs.

segmented backplane A centralized wiring hub that provides for the microsegmentation of a single LAN into two or more addressable segments connected by a filtering or routing component. *See also backplane.*

shared media A reference to the "public" transport wiring used in a network topology on which all devices compete for access to the transmission media and vie for bandwidth and do not transmit over dedicated (or switched) circuits. Examples include Ethernet, FDDI, and Token-Ring with standard concentrator, hub, or MAU technology. *See also shared service.*

shield A barrier, usually metallic, within a wiring bundle that is intended to contain the high-powered broadcast signal within the cable. The shield reduces electromagnetic interference and signal loss.

shielded cable Cable that has a layer of shielded insulation to reduce EMI, RFI, and crosstalk.

shielded twisted-pair Pairs of 22 to 26 AWG wire clad with a metallic signal shield. Abbreviated as STP.

short A physical discontinuity (usually electrical, rarely optical) such that one or more signal conductors in the network media leaks signal into other conductors. Usually, a short refers to an electric short circuit between signal conductor and the shield or grounds. It can also refer to a short between receive or transmit pairs.

short packet A packet that is less than the minimum legal size, including address, length, and CRC fields.

signal A transmission broadcast. The electrical pulse that conveys information.

signal propagation speed The speed at which the signal wave passes through the transmission channel. This is the speed of electrical pulses in copper, light in fiber, or radio signals through air and building materials.

signal quality error A transmission sent by a transceiver back to the Ethernet controller to let the controller know whether the collision circuitry is functional. A background indicator (or "heartbeat") signal that provides a carrier signal. Abbreviated as SQE.

signaling The process of sending a transmission signal over a physical medium for purposes of communication.

skew between pairs The difference between signal arrival times from two initially coincident signals propagated over two different wiring pairs as measured at the receiving end of the cable. This can be a problem with quartet signaling.

sliding window flow control This is a method of flow control in which a receiver gives a transmitter permission to transmit data until a window is full. When the window is full, the transmitter must stop transmitting until the receiver advertises a larger window. TCP, other transport protocols, and several link-layer protocols use this method of flow control.

slot time 1. The time during which the protocol allows a node or station to transmit. 2. A multipurpose parameter to describe the contention behavior of the Data Link Layer. It is defined in Ethernet as the propagation delay of the network for a minimum-size packet (66 bytes). Slot time provides an upper limit on the collision vulnerability of a given transmission, an upper limit on the size of the frame fragment produced by the collision runt frame, and the scheduling time for collision retransmission.

SONET Acronym for Synchronous Optical Network (q.v.).

source address The transmitting station's logical address.

Spanning Tree Algorithm An algorithm, the original version of which was invented by Digital Equipment Corporation, used to prevent bridging loops by creating a spanning tree. The IEEE 802.1d standard that detects and manages logical loops in a network. When multiple paths exist, the bridge or router selects the most efficient one. When a path fails, the tree automatically reconfigures the network with a new active path. The algorithm is now documented in the IEEE 802.1d specification, although the Digital algorithm and the IEEE 802.1d algorithm are not the same, nor are they compatible. Abbreviated as STA.

spoofing The process of sending a false or dummied acknowledgment signal in response to a request for status or receipt. Spoofing is typically applied for host transmission over LAN-type transmission networks or routers so that processes are not falsely terminated for lack of message response activity.

SQE Acronym for signal quality error (q.v.).

STA Acronym for Spanning Tree Algorithm (q.v.).

standard A commonly used or officially specified set of rules or procedures.

star topology A LAN topology in which endpoints on a network are connected to a common central switch by point-to-point links.

StarLAN 1. Another name for 1-Mbps IEEE 802.3. 2. Another name for 1Base5. 3. A 1-Mbps CSMA/CD LAN promulgated by AT&T. A

TCP/IP and Ethernet version that uses twisted-pair or telco wiring in place of coaxial cable. It operates at transmission speeds of 1 Mbps, although 3 Mbps or 10 Mbps are possible.

station 1. A logical, nonphysical interconnection to the network that supports computer workstations or other types of physical devices on a network. Alternatively, a station may connect to a wiring concentrator providing network access for many devices. A device might be a terminal server, or a shared peripheral such as a file server, printer, or plotter. 2. A single addressable device on FDDI, generally implemented as a stand-alone computer or a peripheral device such as a printer or plotter. 3. A station might be a terminal server or a shared peripheral such as a file server, printer, or plotter. *See also node or workstation.*

store-and-forward A message-switching technique where messages are temporarily stored at intermediate points between the source and destination until such time as network resources (such as an unused link) are available for message forwarding.

store-and-forward switch mode The switching technology that has the most latency; its latency is equivalent to that of a router because it is a MAC-layer bridge and packet router. This entails a substantial delay, because arriving packets are fully buffered and checked for errors before they are switched, and when switched they vie for slot time on either the destination node or destination subnet. Note that packets can be switched not only between paired nodes, but also between paired subnets. *See also cut-through switch mode and modified cut-through switch mode.*

STP Acronym for shielded twisted-pair (q.v.).

structured wiring Adherence to the concept of installing a general-purpose and multi-purpose copper (and fiber) plan for telecommunications and data communications. *See also EIA/TIA 568 and EIA/TIA 569.*

subnet For routing purposes, IP networks can be divided into logical subnets by using a subnet mask. Values below those of the mask are valid addresses on the subnet.

subnetwork A term sometimes used to refer to a network segment. In IP networks, a network sharing a particular subnet address. In OSI networks, a collection of ESs and ISs under the control of a single administrative domain and using a single network access protocol.

switched ethernet A technology that provides a switched circuit between paired communicating Ethernet network nodes.

Switched Multimegabit Digital Service Any of a variety of switched digital services ranging from 1.544 megabits to 44.736 megabits per second (T-1 to T-3 speeds). Abbreviated SMDS.

synchronization 1. The process of achieving a common interpretation of a transmitted bit stream between more than one entity at the same point in the bit stream. Synchronization may be required at each layer of a protocol stack. 2. The event occurring when transmitting and receiving stations operate in unison for very efficient (or inefficient) utilization of the communications channel. 3. Establishing common timing between a sender and receiver.

synchronous optical network A common carrier fiber-optic transmission link providing basic bandwidth in blocking units of 45 Mbps. Multiple streams can support bandwidths up to 18 billion bits per second. Used with ATM protocols. Abbreviated as SONET.

TCP/IP Acronym for transaction control protocol/internet protocol. Although commonly referred to as TCP/IP, a complete implementation of this networking protocol includes Transmission Control Protocol (TCP), Internet Protocol (IP), Internetwork Control Message Protocol (ICMP), User Datagram Protocol (UDP), and Address Resolution Protocol (ARP). Standard applications are File Transfer Protocol (FTP), Simple Mail Transfer Protocol (SMTP), and TELNET, which provide virtual terminal on any remote network system. Acronym for Transaction Control Protocol/Internet Protocol (q.v.). *See also TCP and IP.*

technical service bulletin An addendum to an ANSI/EIA/TIA specification. Abbreviated as TSB.

Teflon Registered trade name for fluorinated ethylene propylene. A nonflammable material used for cable foam and jacketing.

telecommunications A term referring to communications (usually involving computer systems) over the telephone network.

throughput 1. A measurement of work accomplished. 2. The volume of traffic that passes through a pathway or intersection. Typically, this refers to data communications packets or cells and it is measured in packets/s, cells/s, or bits/s. 3. Rate of information arriving at, and possibly passing through, a particular point in a network system.

TIA *See Electronics Industry Association/Telecommunication Industry Association.*

time-out 1. An event that occurs when one network device expects to hear from, but does not hear from, another network device within a specified period of time. The resulting time-out usually results in a retransmission of information or the outright dissolving of the virtual circuit between the two devices. 2. The explicit failure to receive a status message during a synchronization window, or the implicity failure to receive an acknowledgment message, both of which result in session failure. 3. A method to train a toddler as to appropriate behavior.

token 1. The protocol-based permission that is granted to a station in a predetermined sequence. The permission allows that station to transmit on the network. 2. A control information frame, possession of which grants a network device the right to transmit.

token passing An access method by which network devices access the physical medium in an orderly fashion based on possession of a small frame called a token. *See also contention and circuit switching.*

token ring A physical networking configuration.

Token-Ring 1. An IBM network protocol and trademark. 2. A popular example of a local area network; a precursor of the IEEE 802.5 standard, which was derived from original IBM working papers. Token-Ring applies the IEEE 802.2 MAC protocols and uses the nonpersistent token protocol on a logical ring (although a physical star) topology. Transmission rate is 4 Mbps, with upgrades to 16 Mbps and options to release a token upon completion of frame transmission (early token release), and burst mode option. *See also token ring.*

topology Layout of a network. This describes how the nodes are physically joined to each other. *See also bus, ring, and star.*

traffic 1. The communications carried by a system. 2. A measure of network load that refers to the frame transmission rate (frames per second or frames per hour).

traffic congestion The situation when data arrivals exceed delivery times and create performance problems.

traffic jam A metaphorical reference to congestion on a network.

Transaction Control Protocol/Internet Protocol *See TCP/IP.*

transceiver The coaxial cable interface based upon a mechanical and electrical connection to the Ethernet coaxial cable medium. It is typically a radio transmitter and receiver combined to form a single unit. A cellular phone uses a transceiver to send signals to, and receive them from, the cell site. Also known as a medium-dependent interface.

transceiver cable A four-pair shielded cable that interconnects a workstation to a transceiver or fan-out box. Often called a drop cable or AUI cable.

transmission Any electronic or optical signal used for telecommunications or data communications to send a message.

TSB Acronym for technical service bulletin (q.v.).

tunneling Encapsulation. The process of enclosing one protocol inside another protocol packet for delivery over a router-defined link.

twisted pair Telephone wire twisted over its length to preserve signal strength and minimize electromagnetic interference. Abbreviated as TP.

Type 1 A dual-pair, 22 AWG cable with solid conductors and a braided shield. It is a type of shielded twisted-pair. Type 1 cable corresponds to TIA/EIA Category 4 and possibly Category 5 performance recommendations.

Type 1 operation IEEE 802.2 (LLC) connectionless operation.

Type 2 A six-pair, shielded, 22 AWG wire used for voice transmission. It is the same wire as Type 1, but it has an additional four-pair wire for telephone services.

Type 3 A single-pair, 22 or 24 AWG, unshielded twisted-pair wire. It is common telephone wire.

Type 5 IBM 100/140-micron fiber; IBM now recommends 125-micron fiber.

Type 6 A two-pair, stranded, 26 AWG wire used for short patch cables.

Type 8 A two-pair, 26 AWG, shielded cable without any twists; it is commonly used under carpet. Also called silver satin.

type error A packet that is improperly labeled with protocol information.

Ultra Fast Ethernet Any reference to the proposed 1-Gbps enhancement to Ethernet.

undersized error A frame or packet that does not meet the minimum size requirement. Oversized packets indicate that a faulty or corrupted driver is in use on the network.

undersized packet A packet that contains less than 64 bytes, including address, length, and CRC fields.

unshielded twisted pair Pairs of 22 to 26 AWG wire usually in bundles of 2, 4, or 25 pairs installed for telephone service and occasionally for data networks. Referred to as voice-grade twisted-pair or voice-grade wiring. Abbreviated as UTP.

UTP Acronym for unshielded twisted pair (q.v.).

virtual circuit A logical circuit set up to ensure reliable communication between two network devices.

virtual path A group of virtual channels that can support multiple virtual circuits.

voice grade EIA/TIA category 1, 2, or 3 twisted-pair wiring best used for analog lines rather than high-speed data communication lines. *See Twisted-Pair.*

WAN Acronym for wide area network (q.v.).

wide area network A network that spans cities, states, countries, or oceans. Command carrier services usually supply the network links. Abbreviated WAN.

wiring closet A specially designed room used for wiring data and voice networks. Wiring closets serve as central junction points for wiring and wiring equipment that is used for interconnecting devices.

wiring concentrator A central wiring concentrator for a series of Ethernet, FDDI, and Token-Ring nodes. *See also wiring hub.*

wiring hub A central wiring concentrator for a series of Ethernet, FDDI, and Token-Ring nodes.

workstation 1. Any computer device. 2. Any device on a network. 3. A single addressable site on FDDI that is generally implemented as a stand-alone computer or a peripheral device, connected to the ring with a controller. *See also node or station.*

SOURCES

A 'n D Cable Products, Concord, CA.

American National Standards Institute (ANSI). *Advanced Data Communications Control Procedures (ADCCP).* ANSI X3.66. New York, 1979.

DEC-Intel-Xerox. *The Ethernet: A Local Computer Network, Data Link Layer and Physical Layer, Version 1.0.* September 30, 1980.

Fetterolf, James. AMP Inc., Valley Forge, Pennsylvania 01776.

Folts, Harold C. editor. *Data Communications Standards, Third Edition.* McGraw-Hill, 1988.

Hiller, Frederich S., and Lieberman, Gerald J. *Operations Research.* Holden-Day, Inc., 1974.

Leclerc, Rick. Software Project Leader, Cabletron Systems. Rochester, NH.

Matibag, Joseph. Director of Marketing, Wavetek Corporation. San Diego, CA 92123.

National Security Decisions Directive (September 17, 1984, #14 unclassified).

Network Performance Institute, Inc., Miami Beach, FL.

Peaslee, George. Field Support Technician, Racal-Datacom. Sunrise, FL.

Shannon, C. E. *The Mathematical Theory of Communications.* University of Illinois Press, Urbana, Illinois, 1964.

Shoch, John F., and Hupp, Jon A. *Performance of an Ethernet Local Network—A Preliminary Report.* Local Area Communications Symposium. Mitre and NBS. Boston Massachusetts, May 1979.

SRI International. *Internet Protocol Transition Workbook.* Menlo Park, California 94025. March, 1981.

Wang, P. T., and McGinn, Michael. *Mitre Corporation: Performance of a Stochastically Optimized CMSA Network.* /0742/1303/85/000/0061 IEEE. 1979.

INDEX

D

E

R

S

W